FEEDBACK MAXIMIZATION

The Artech House Microwave Library

Introduction to Microwaves by Fred E. Gardiol
Microwaves Made Simple by W. Stephen Cheung and Frederic Levien
Microwave Tubes by A. Scott Gilmour, Jr.
Electric Filters by Martin Hasler and Jacques Neirynck
Microwave Technology by Erich Pehl
Receiving Systems Design by Stephen J. Erst
Applications of GaAs MESFETs by R.A. Soares, et al.
GaAs Processing Techniques by R.E. Williams
GaAs FET Principles and Technology by J.V. DiLorenzo and D. Khandelwal
Dielectric Resonators, Darko Kajfez and Pierre Guillon, eds.
Modern Spectrum Analyzer Theory and Applications by Morris Engelson
Microwave Materials and Fabrication Techniques by Thomas S. Laverghetta
Handbook of Microwave Testing by Thomas S. Laverghetta
Microwave Measurements and Techniques by Thomas S. Laverghetta
Principles of Electromagnetic Compatibility by Bernhard E. Keiser
Design and Analysis of Linear Active Circuits by William Rynone, Jr.
The Design of Impedance-Matching Networks for Radio-Frequency and Microwave Amplifiers by P.L.D. Abrie
Microwave Filters, Impedance Matching Networks and Coupling Structures by G.L. Matthaei, Leo Young, and E.M.T. Jones
Microwave Engineer's Handbook, 2 vol., Theodore Saad, ed.
Handbook of Microwave Integrated Circuits by R.K. Hoffmann
Microwave Integrated Circuits, Jeffrey Frey and Kul Bhasin, eds.
Computer-Aided Design of Microwave Circuits by K.C. Gupta, R. Garg, and R. Chadha
Microstrip Lines and Slotlines by K.C. Gupta, R. Garg, and I.J. Bahl
Microstrip Antennas by I.J. Bahl and P. Bhartia
Antenna Design Using Personal Computers by David M. Pozar
Microwave Circuit Design Using Programmable Calculators by J. Lamar Allen and Max Medley, Jr.
Stripline Circuit Design by Harlan Howe, Jr.
Microwave Transmission Line Filters by J.A.G. Malherbe
Electrical Characteristics of Transmission Lines by W. Hilberg
Multiconductor Transmission Line Analysis by Sidney Frankel
Microwave Diode Control Devices by Robert V. Garver
Tables for Active Filter Design by Mario Biey and Amedeo Premoli
Active Filter Design by A.B. Williams
Laser Applications by W.V. Smith
Ferrite Control Components, 2 vol., Lawrence Whicker, ed.
Microwave Remote Sensing, 3 vol., by F.T. Ulaby, R.K. Moore, and A.K. Fung

FEEDBACK MAXIMIZATION

B. J. LURIE

Jet Propulsion Laboratory
California Institute of Technology

Copyright © 1986

ARTECH HOUSE, INC.
610 Washington Street
Dedham, MA 02026

All rights reserved. Printed and bound in the United States of America. No part of this book may be reproduced or utilized in any form or by any means, electronic or mechanical, including photocopying, recording, or by any information storage and retrieval system, without permission in writing from the publisher.

International Standard Book Number: 0-89006-200-5
Library of Congress Catalog Card Number: 86-70365

86 87 10 9 8 7 6 5 4 3 2 1

Contents

PREFACE		x
INTRODUCTION — NONLINEAR DYNAMIC COMPENSATION		xi
CHAPTER 1 LINEAR SINGLE-LOOP FEEDBACK SYSTEM		1
1.1	Transmission and Driving-Point Impedance	1
	1.1.1 Definitions	1
	1.1.2 Generic Single-Loop System	3
	1.1.3 Inverse Block Diagram; Two-Poles' Connection	10
	1.1.4 Return Ratio as a Function of the Load	13
	1.1.5 Balanced Bridge	14
	1.1.6 Examples and Exercises	16
1.2	Sensitivity	24
	1.2.1 Sensitivity of Transmission Function	24
	1.2.2 Sensitivity of Driving-Point Impedance	26
	1.2.3 Reflection Coefficients	27
	1.2.4 Examples and Exercises	28
1.3	Nonlinear Distortion	28
1.4	Regulation	29
	1.4.1 Introduction	29
	1.4.2 Parameter Dependence of a Circuit Function	29
	1.4.3 Symmetrical Regulation	32
1.5	Noise	35
	1.5.1 Noise at the System's Output	35
	1.5.2 Noise at the Input of the Plant	36
	1.5.3 Signal-to-Noise Ratio	36
	1.5.4 Examples and Exercises	38
CHAPTER 2 STABILITY AND FREQUENCY RESPONSE CONSTRAINTS OF LINEAR SYSTEMS		45
2.1	Nyquist Stability Criterion	45
	2.1.1 Nyquist Diagram	45
	2.1.2 Nyquist Diagram and Stability Margins for the Amplifier Return Ratio	48
	2.1.3 Two-Poles' Coupling	52
	2.1.4 Reflection Coefficients	52
	2.1.5 Examples and Exercises	54

2.2	Integral Constraints	61
	2.2.1 Minimum Phase Functions	61
	2.2.2 Integral of the Real Part	61
	2.2.3 Integral of the Imaginary Part	68
	2.2.4 Examples and Exercises	69
2.3	Phase-Gain Relations	72
	2.3.1 General Relation	72
	2.3.2 Phase Response Calculation	75
	2.3.3 Piecewise-Constant Real and Imaginary Components	77
	2.3.4 Examples and Exercises	84
2.4	Feedback Maximization	88
	2.4.1 Bode Optimal Cut-Off	88
	2.4.2 More Cut-Offs	94
	2.4.3 Asymptotic Losses, High-Frequency Bypass, Loop Gain Correction in Feedback Amplifiers	102
	2.4.4 Prediction and Feedback Maximization	104
	2.4.5 Negative Resistance Sensitivity	106
	2.4.6 Examples and Exercises	107
2.5	Nonminimum Phase Shift (NPS)	113
	2.5.1 Causes for the Nonminimum Phase Shift	113
	2.5.2 Parallel Connection of Two Links	114
	2.5.3 Effect of Loading	119
	2.5.4 Noncascade Connection of Active Two-Ports	123

CHAPTER 3 SYSTEM WITH SINGLE NONLINEAR LINK … 129

3.1	Absolute Stability Problem	129
3.2	Popov Criterion	130
	3.2.1 Nonlinear Physical Two-Poles	130
	3.2.2 Popov Criterion	133
	3.2.3 Applications	137
	3.2.4 Examples and Exercises	139
3.3	Periodical and Nonperiodical Self-Oscillation	142
3.4	Multifrequency Oscillation in a Bandpass System with Saturation	143
	3.4.1 Goals for Analysis	143
	3.4.2 Oscillation with Fundamental at which the Loop Gain is Large	143
	3.4.3 Oscillation with Fundamental at which the Loop Gain is Small	146
3.5	Describing Function (DF) Approach	152

3.6		Describing Functions for the Basic Types of Nonlinear Links	154
3.7		Multivalued Output-Input Relations	159
	3.7.1	Two-Pole with Negative dc Resistance	159
	3.7.2	Two-Port	162

CHAPTER 4 FORCED OSCILLATION ... 165

4.1		Periodic Excitation	165
4.2		Absolute Stability of the Output Process	170
4.3		Jump-Resonance	176
	4.3.1	Conditions for the Jumps	176
	4.3.2	System with Dynamic Saturation	179
	4.3.3	System with Nondynamic Nonlinear Link	182
	4.3.4	System with Nondynamic Saturation	183
	4.3.5	Substantiation of DF Technique for the Jump-Resonance Analysis	185
	4.3.6	System with Dead-Zone Element	186
	4.3.7	System with Nonlinear Element Having Power-Type DF	187
	4.3.8	Examples and Exercises	188
4.4		Odd Subharmonics	189
4.5		Second Subharmonic	191

CHAPTER 5 NONLINEAR DYNAMIC COMPENSATOR (NDC) ... 197

5.1		Loop Regulation	197
	5.1.1	Variable Loop Gain	197
	5.1.2	Switching in the Compensator	199
	5.1.3	Quasilinear Variable Compensator	202
	5.1.4	Nonlinear Dynamic Compensator	203
5.2		Describing Function Approach to NDC Design	205
	5.2.1	Generalities	205
	5.2.2	Stability Conditions	206
	5.2.3	Phase Stability Margin	208
	5.2.4	Design Constraints	208
	5.2.5	Amplitude Characteristics	215
	5.2.6	Effect of Loop Gain Ignorance	217
	5.2.7	Sufficient Stability Criterion	218
5.3		NDC in the Interstage Circuit of a Wideband Feedback Amplifier	221
5.4		Experiments	227
5.5		Specifical Applications	231

CHAPTER 6 LINEAR MULTILOOP SYSTEM 233
6.1 Generalities 233
6.2 Stability Criteria 234
    6.2.1 Generalization of the Nyquist Criterion 234
    6.2.2 Examples and Exercises 236
6.3 Feed-Forward 237
6.4 System with Parallel Amplification Channels 242
    6.4.1 Sensitivity 242
    6.4.2 Stability 243

CHAPTER 7 NONLINEAR MULTILOOP SYSTEM:
DESCRIBING FUNCTION APPROACH 249
7.1 Local Feedback 249
    7.1.1 Cascaded Links 249
    7.1.2 Feedback around the Ultimate Stage 249
7.2 Two Parallel Channels with Saturation 253
    7.2.1 Frequency Responses of the Linear Part 253
    7.2.2 Piecewise Analysis of the AAPC 256
    7.2.3 Jump-Resonance 260
    7.2.4 Frequency Response of the Main Channel Transmission Function 261
    7.2.5 Modifications 261
    7.2.6 Numerical Examples 265
    7.2.7 Experiment 267
7.3 Two Parallel Channels with Saturation and Dead Zone ... 269
    7.3.1 Frequency Responses of the Linear Part 269
    7.3.2 Jump-Resonance 271
    7.3.3 Nonlinear Dynamic Compensator 272
    7.3.4 Experiment 272

CHAPTER 8 NONLINEAR MULTILOOP SYSTEM:
ABSOLUTE STABILITY (AS) APPROACH 275
8.1 System Reducible to Single-Channel 275
    8.1.1 Block Diagrams 275
    8.1.2 Integral Constraint and Stability Margin 278
    8.1.3 Plant with Saturation 279
8.2 Dead-Zone Element in the NDC Feedback Path 280
8.3 Positive and Negative Feedback in the NDC 286
8.4 DF *versus* AS Techniques 287

8.5	\multicolumn{2}{l}{Two-Channel System with a Lowpass in the Plant}	289
	8.5.1	Block Diagram	289
	8.5.2	Noise	294
	8.5.3	Other Points of View	294
	8.5.4	The Allowable Discrepancy of Nonlinear Link Characteristics	296
	8.5.5	Experiment	299
8.6	\multicolumn{2}{l}{NDC with Two Nonlinear Elements}	304
8.7	\multicolumn{2}{l}{Switching Regulation}	306
LIST OF SYMBOLS			309
BIBLIOGRAPHY			313
INDEX			325

Preface

This book describes synthesis methods for feedback systems with nonlinear dynamic compensation, which allows for increased feedback while preserving robustness, global stability, good transient responses, and, if required, stability of the output processes. The loop transmission functions are found under the control of Bode's integral relationships.

The text is intended for the designers of single-loop and multiloop feedback systems employed in controllers, tracking systems, amplifiers, phase-locked loops, active filters, power supplies, *et cetera*. The book covers the material needed for realization of practical design goals from the areas of linear systems, nonlinear oscillation, and stability of nonlinear systems.

The three main parts of the book are:

Linear feedback systems: General relationships, integral constraints, Bode's theory of feedback maximization (Chapters 1, 2, and 5). This material is based on the classical works of H.W. Bode, but the whole system of exposition, and proofs of certain theorems have been adapted to the task of the book. New topics are added, such as several theorems about non-minimal phase shift and feedback systems with parallel forward channels.

Nonlinear feedback system analysis: fundamentals of the frequency-domain approach (Chapters 3 and 4) integrating traditional items with newer results, such as a simplified proof of the Popov criterion, necessary conditions for the absolute stability of output processes, calculation of the jumps of jump-resonance, analysis of subharmonics and multifrequency oscillation.

Synthesis methods for globally-stable Nyquist-stable feedback systems with increased feedback (Chapters 5, 7, and 8): The synthesis procedures based on the absolute stability requirements and on the describing function technique have been applied to the design of single-loop and multiloop systems with nonlinear dynamic compensation. In numerical examples, as well as in experimental devices and industrial feedback systems, substantial improvements in the sensitivity and the closed-loop transient response have been demonstrated over those of systems constructed with conventional methods.

Courses based on this text were taught at the Weizmann Institute of Science and Tel-Aviv University. In a letter written in 1980, H.W. Bode commented on the author's papers for the control system course: "With a few additions, perhaps from other sources, including my past work, the papers make a very interesting and provocative basis for systematic treatment on a large scale."

Readers who are interested in automatic control only (and not in feedback amplifiers) may skip subsections 1.1.2, 1.1.4–1.1.5, 1.2.2–1.2.4, 1.3, 2.4.3, 2.5.3, 2.5.4, and 5.3.

INTRODUCTION
Nonlinear Dynamic Compensation

Given sufficient feedback, system operation becomes precision and almost independent of the parameters of the actuator and the object of regulation, both of which normally have extensive parameter uncertainty. However, the attainable feedback for the prescribed speed of control processes (in other words, over the operational frequency band) is invariably constrained, as will be seen in the following.

At least one of several factors very definitely limits the frequency at which the gain around the loop crosses the 0 dB level. Among these factors are overloading the actuator by high-frequency noise components of the sensors and amplifiers, degrading gain of the employed amplifiers at higher frequencies, resonances in flexible mechanical structures, delays of the signal propagation around the control loop, and delays of digitizing and signal processing in computers. Hence, the available feedback depends on the steepness of the slope of the loop gain-frequency response in the frequency interval between the operational band and the crossover frequency. This slope, in turn, is linked by the Bode functional with the loop phase lag, the latter restricted by stability conditions. Consequently, the value of the feedback is limited.*

Assuming the effects of saturation in the actuator may be adequately approximated by reduction of the loop gain, H. W. Bode found the frequency responses of the feedback loop transmission that maximize the feedback $|F|$ over the operational frequency band. Accordingly, the phase lag in the loop was bounded by $(1 - y)180°$ at the frequencies where the loop gain exceeds $-x$ dB. Typical values for the phase stability margin denoted as $y180°$ are 30° to 50°, and for the amplitude stability margin x, of 10 dB. The product *bandwidth* $\times |F|^{0.7}$ is constrained by the location of the crossover frequency, thus establishing a simple rule for the design trade-off between the regulation accuracy, i.e., the value of the feedback, and the maximal speed of the signal variations at which this accuracy is still maintained.

---

*This constraint might not be seen when a system is idealized as linear and designed for minimal error at the system output. In physical implementation of such a system, the high frequency noise, negligible at the system's output due to the object of regulation inertia, may overload the actuator; and the actuator nonlinearity may cause extensive process instability and oscillation.

The optimal loop gain response was found to be like that of a lowpass filter, with the cut-off slope close to –10 dB/octave. This steep slope substantiates neglecting the harmonics of conceivable oscillation in the stability analysis, i.e., using describing function approach for proving the system asymptotic global stability (AGS). Later, it was confirmed with the methods of absolute stability (AS) that the system with the optimal Bode cut-off is AGS when the loop nonlinear link is memoryless, described by the function $v(e)$, where $0 < v/e < 1$.

The slope of the loop gain response may be further increased in a Nyquist-stable system, where the loop phase lag is allowed to exceed 180° at all frequencies except in the vicinity of the crossover frequency. Consequently, much higher feedback may be attained. This feature is indeed of fundamental importance.

Nyquist-stable systems do not satisfy the absolute stability criterion. If such a system contains a nonlinear memoryless link, its stability appears to be conditional (Liapunov called a system conditionally stable if initial conditions from a certain set lead to stability, while other conditions lead to instability). To ensure unconditional asymptotic global stability, a nonlinear dynamic compensator (NDC) must be inserted into the loop.

A design procedure for the synthesis of a feasible, generic nonlinear dynamic link is yet to be developed. Fortunately, the problem dramatically simplifies after shrinking the applicability area to only the systems of practical interest, as diagramed in Fig. I.1.

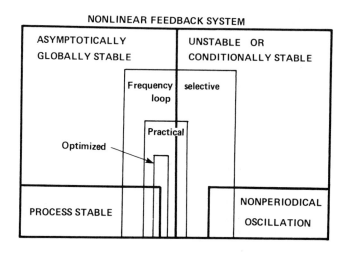

FIG. I.1

Systems with maximized feedback are neither the low-order systems (which would otherwise substantiate a differential equations approach), nor the nearly linear systems (which would make it reasonable to start with well developed linear system methods), but high-order nonlinear systems. Their loop phase lags are large (otherwise, extra phase margins may be traded off for extra feedback), and the loop gain responses are like those of a filter. These specific responses naturally call for frequency-domain design methods, either describing functions (DF) or the methods of absolute stability.

In the process of design and local optimization, the engineer must deal with systems that are very nearly optimal. Self-oscillation in such systems is, as a rule, periodic, and, due to filtering in the linear links, nearly sinusoidal at the input to the nonlinear links. This allows for employing a describing function stability analysis. The phase shift errors of the analysis is less than 15° in practical examples, which is small compared to the phase lead of up to 200° produced by the NDC.

In nonlinear system operation, certain restrictions must be imposed on the transients of the output signal following the overloading of the actuator by sporadically large input signals. The system may be characterized with either the step test signal in the time domain by the value of wind-up (long overshoot), or in the frequency domain by the value of jump-resonance, i.e., hysteresis-like jumps in the amplitude and phase of the output following gradual changes in the amplitude of the input sinusoidal signal.

These nonlinear phenomena could be eliminated by making the system process stable. The process stability in systems with a single nondynamic nonlinear link is, however, only achievable at the price of a higher phase stability margin, thus dramatically reducing the available feedback. Therefore, process stability is rarely preserved, and the forced oscillation contributes to errors in control systems and feedback amplifiers.

Utilizing the NDC excludes the forced oscillation without sacrificing the available feedback. This yields the second benefit of NDC applications.

Although this book concentrates on single-input, single-output multiloop system design, nonlinear dynamic compensation is recommended for application in multi-input, multi-output systems as well. Here, for the small price of the compensators at the inputs to some of the actuators, the phase stability margin will smoothly increase as the amplitude of the signal grows over the saturation threshold, and the interaction between the loops caused by unavoidable, otherwise positive, nonlinear feedback will vanish. These considerations might allow the designer to confine the stability analysis to the linear mode of operation, and to challenge the design of multidimensional integrated systems.

Chapter 1
Linear Single-Loop Feedback System

1.1 TRANSMISSION AND DRIVING-POINT IMPEDANCE

1.1.1 Definitions

The block-diagram for the linear single-loop feedback system is displayed in Fig. 1.1. The forward-path transfer function is $K(j\omega)$; the transfer function of the feedback path, $\beta(j\omega)$.

The signal E at the input to the two-port K causes the *return signal* $-K\beta E$ to be fed back to the combiner. The ratio of these signals taken with the opposite sign gives the *return ratio* $T = K\beta$.

The signal U_1 applied to the input of the system yields $E = U_1 - ET$ so that

$$E = U_1/(T+1) = U_1/F \tag{1.1}$$

In the following, $F = 1 + K\beta = 1 + T$ is referred to as the *return difference*; $|F|$ in dB, or merely $|F|$ as the *feedback*; and $|T|$ in dB, as the *loop gain*.

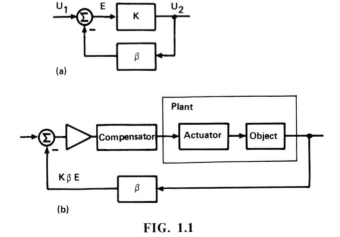

FIG. 1.1

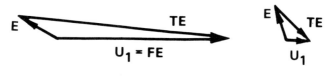

FIG. 1.2

The system is said to have *positive feedback* if application of the feedback increases the input and, consequently, the output signals of the forward path. Positive feedback produces a positive increment in dB in the output signal. From (1.1), the condition for positive feedback is $|F|<1$. When $|F|>1$, the feedback is said to be *negative*. The phasor diagrams of Fig. 1.2 illustrate performing (1.1) for negative and positive feedback.

The *closed-loop transfer function* is seen as

$$K_\beta = \frac{U_2}{U_1} = \frac{K}{F} = \frac{K}{K\beta+1} = \frac{1}{\beta}\frac{T}{F} = \frac{1}{\beta}M \tag{1.2}$$

where M is expressed in T and F as

$$M = \frac{T}{F} = \frac{1}{1+1/T} = 1 - \frac{1}{F}$$

If $K \to \infty$, then $F \to \infty$, $M \to 1$ and

$$\lim_{K \to \infty} K_\beta = \frac{1}{\beta} \tag{1.3}$$

The closed-loop transfer function can, therefore, be expressed as

$$K_\beta = \left(1 - \frac{1}{F}\right) \lim_{K \to \infty} K_\beta \tag{1.4}$$

K_β approaches the value of (1.3) with the increase of feedback.

If the gain of the amplifier in Fig. 1.1(b) is sufficiently large, the relations of *large feedback* follow:

$$\left.\begin{array}{r}|F| \gg 1 \\ F \cong T \\ M \cong 1 \\ K_\beta \cong \dfrac{1}{\beta}\end{array}\right\} \tag{1.5}$$

The large feedback makes K_β dependent exclusively on the β-path. Most frequently $|K_\beta| > 1$, and the output of the feedback path $|E/F|$ is significantly smaller than the system output. Given this, the β-path can be constructed of only passive components that can be made precision.

This is of fundamental importance because the forward path normally includes links with a much wider range of parameter deviations (i.e., having large ignorance [72]). Typical examples of these links are the power supplying element (actuator) and the device under control, which together constitute what is called the plant, as indicated in Fig. 1.1(b).

When, for example, K varies within the range from K' to K'', the relative change in the closed loop transmission coefficient is

$$\frac{K'_\beta}{K''_\beta} = \frac{M'}{M''} = \frac{1 - 1/F'}{1 - 1/F''} \tag{1.6}$$

If the feedback remains large with both K' and K'', this ratio is approximated by

$$\frac{K'_\beta}{K''_\beta} = 1 - 1/F' + 1/F'' \cong 1 - \left(1 - \frac{K'}{K''}\right)\frac{1}{F'} \tag{1.7}$$

With the frequency hodograph of $T(j\omega)$ plotted onto the Nichols chart shown in Fig. 1.3, $|M|$ in dB and arg M can be read from the curvilinear coordinates of the chart. The chart also serves to determine F from a known T. With the locus for $1/T(j\omega)$ plotted, the curvilinear coordinates give $1/F(j\omega)$.

A system with large feedback and unity transmission of the feedback path, $\beta = 1$, forms a tracking system, where the output tracks the input. The accuracy of the tracking depends on the bandwidth of the frequency range where the feedback is sufficiently large. The broader the bandwidth, the faster input signals can be tracked with the prescribed accuracy.

The maximum bandwidth of large feedback is limited, as will be demonstrated in the next chapter. It will be also shown that linear or nonlinear dynamic compensators must be inserted into the loop to provide stability. These links are normally installed into the forward path at the input to the plant so as not to affect the external gain of the system, as illustrated in Fig. 1.1.(b).

1.1.2 Generic Single-Loop System

The previous analysis is generalized in this section for a system comprising an arbitrary passive network β as displayed in Fig. 1.4.

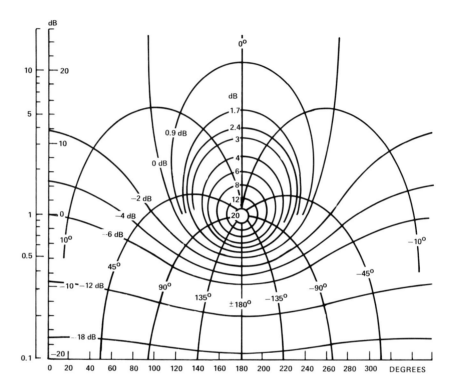

FIG. 1.3

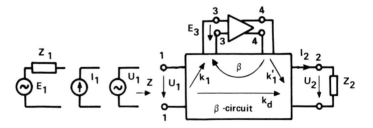

FIG. 1.4

The active two-port is assumed to have zero reverse transmission and either infinite or infinitesimal input and output impedances. To apply this structure to the analysis of systems with physical amplifiers, the amplifier input and output impedances may be viewed as produced by passive two-poles connected in parallel or series to the amplifier input and ouput. Further, these two-poles must be integrated into the β-network.

The dimensionality of the signals at the input and output of the amplifier, and at the input and output terminals of the whole system, in principle, does not influence the following analysis. However, to simplify the exposition, we will select one of the possible versions and characterize the signal at the input to the active element by the voltage E_3, and the signal at its output by the current I_4. The active element with transadmittance $w_{43} = I_4/E_3$ is, therefore, assumed to have high input and output impedances.

When an external electromotive force (e.m.f.) E_3 is applied to the input of the amplifier disconnected from the β-circuit, as shown in Fig. 1.5, the return voltage U_3 appears between the poles $3'$-$3'$. The return ratio is

$$T = -\frac{U_3}{E_3} \tag{1.8}$$

When measuring the return ratio T, the e.m.f. E_1 of the source connected to the system's input 1-1 must be replaced by a short circuit, in accordance with the superposition principle. Thus, the source impedance Z_1 is connected to the poles 1-1. The return ratio, therefore, depends on Z_1, i.e., $T = T(Z_1)$. In particular $T(0)$ denotes the value of T measured while the input is shorted, and $T(\infty)$, while the input is open.

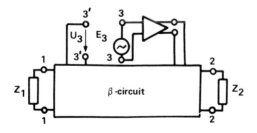

FIG. 1.5

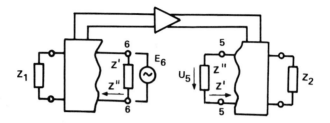

FIG. 1.6

Figure 1.6 shows the *cross-sectioned* feedback circuit. The external two-poles Z' and Z'' connected to the input and output of the broken loop provide appropriate loading for the disconnected parts of the β-circuit so that their transfer coefficient remain unchanged. The e.m.f. E_6 applied to the input of the broken loop produces return signal U_5. The ratio

$$-\frac{U_5}{E_6} = -\frac{U_3}{E_6}\frac{I_4}{U_3}\frac{U_5}{I_4}$$

equals T, as well as the ratio of (1.7). While measuring T this way, the two-pole Z' need not be connected, since it is shunted by the e.m.f. E_6.

Generally, the voltage output-input ratio of a linear four-pole can always be presented as the product of two ratios: the current output-input ratio, and the load impedance to the four-pole input impedance ratio. While calculating T, the latter ratio equals 1. Therefore, T may be measured arbitrarily as the ratio of either voltages or currents.

In the closed-loop system, the signal U_3 is formed by superposition of the effects U_3^0 and $-U_3 T$, correspondingly produced by the signal source and the output of the amplifier. Thus, $U_3 = U_3^0 - U_3 T$, from where we have

$$U_3 = \frac{U_3^0}{F} \tag{1.9}$$

The output of the feedback system is also a linear combination of two signal sources: the output of the amplifier and the signal source. By virtue of (1.9), the signal from the amplifier output is reduced F times. Hence, the closed-loop system transfer coefficients in voltage, current, and as the ratio of the output voltage to the signal e.m.f. are, respectively,

$$K_\beta = \frac{U_2}{U_1} = \frac{K}{F(0)} + k_d \tag{1.10}$$

$$K_{I\beta} = \frac{I_2}{I_1} = \frac{K_I}{F(\infty)} + k_{Id} \tag{1.11}$$

$$K_{E\beta} = \frac{U_2}{E_1} = \frac{K_E}{F} + k_{Ed} \tag{1.12}$$

where, K, K_I, and K_E are the open-loop system transmission functions, measured while the feedback path is disconnected. $F(0)$, $F(\infty)$, and $F = F(Z_1)$ are the return differences measured under the conditions of the connecting zero impedance, infinite impedance, or impedance Z_1 to the system's input terminals respectively. The coefficients of direct signal propagation through the β-circuit are k_d, k_{Id}, and k_{Ed}, and are determined under the same set of loading conditions at the input terminals.

Let Z designate the input impedance, and Z_0 the input impedance in the system without feedback (with cross-sectioned feedback path, or with $w_{43} = 0$, i.e., with the active element shut off). Then,

$$K_\beta = K_{I\beta} \frac{Z_2}{Z} \tag{1.13}$$

$$k_d = k_{Id} \frac{Z_2}{Z} \tag{1.14}$$

$$K = K_I \frac{Z_2}{Z_0} \tag{1.15}$$

By substituting (1.14) and (1.15) into (1.10), and (1.11) into (1.13), we get

$$\frac{K_I Z_2 / Z_0}{F(0)} + k_{Id} \frac{Z_2}{Z} = \left(\frac{K_I}{F(\infty)} + k_{Id} \right) \frac{Z_2}{Z}$$

from which the formula of Blackman [17,23] follows:

$$Z = Z_0 \frac{F(0)}{F(\infty)} \tag{1.16}$$

The formula expresses Z through three functions, each of which is simple to calculate, as will be demonstrated in the examples, Sec. 1.1.6.

Since any two nodes of the β-circuit, in principle, can be regarded as input terminals of the feedback system, formula (1.16) can be used for calculation of the driving-point impedance between any two terminals n-n', provided that $F(0)$ is understood as F measured with the terminals n-n' shorted, and $F(\infty)$, with these terminals open.

If the terminals are shorted, the voltage between them vanishes, but not the current. For this reason, the feedback is called *current-mode* (or *series*) with respect to the terminals n-n', if $T(0) \neq 0$ and $T(\infty) = 0$ with respect to these terminals. Analogously, the feedback is *voltage-mode* (or *parallel*) if $T(0) = 0$ and $T(\infty) \neq 0$. If neither of them equals 0, the feedback is called *compound*.

When the β-circuit is reciprocal, β is expressed as a transimpedance and the amplifier is characterized by its transadmittance; then $\beta(0)$ and $\beta(\infty)$ do not vary when reversing their direction of transmission. Therefore, reversing the direction of the amplifier does not alter $T(0)$, $T(\infty)$, and, consequently, Z.

If the feedback is infinite, the input impedance is

$$Z_\infty = \lim_{w_{43} \to \infty} Z = Z_0 \frac{T(0)}{T(\infty)} = Z_0 \frac{\beta(0)}{\beta(\infty)} \tag{1.17}$$

depending exclusively on the β-circuit and not on w_{43}.

Combining (1.16) and (1.17) gives $Z = Z_\infty M(\infty)/M(0)$.

According to Fig. 1.4, $K = k_1 k_1' w_{43}$; where k_1 is the transmission coefficient (or, depending on the chosen signal dimensionality, transimpedance or transadmittance) from the input source to the input of the amplifier, and k_1', for the transfer function from the output of the amplifier to the system's output. Thus, from (1.10), it follows that:

$$K_\beta = - \frac{k_1 k_1'}{\beta} M(0) + k_d \tag{1.18}$$

$$\lim_{w_{43} \to \infty} K_\beta = - \frac{k_1 k_1'}{\beta} + k_d \tag{1.19}$$

It can now be proved that both Z_∞ and $\lim (w_{43} \to \infty) K_\beta$ do not depend on the impedance w of a two-pole connected in parallel to the amplifier's input or output, i.e., on the input and output impedances of a physical amplifier.

According to (1.16), if $w_{43} \to \infty$, the driving-point impedance faced by w is infinitesimal. Then, it can be seen from (1.47), which is recruited from Sec. 1.4.2, that W, which may designate either K_β or Z, is independent of w.

We conclude that if the feedback is sufficiently large, the external parameters of a feedback system are practically independent of the parameters of the real amplifier employed in place of the ideal one.

Linear Single-Loop Feedback Systems

Consider next the block diagram of Fig. 1.7, consisting of the input and output six-poles, the amplifier w_{43}, and the four-pole β_0. If the feedback is large, such a system possesses the important feature that its input impedance is exclusively dependent on the parameters of the input six-pole (combiner), and its output impedance, on the parameters of the output six-pole (splitter).

To clarify this statement, the block-diagram is redrawn in Fig. 1.8 such that all subcircuits except the six-pole are integrated in an equivalent amplifier for which the six-pole represents an equivalent feedback circuit. Thus, $Z_\infty = Z_0 k_2(0)/k_2(\infty)$.

In particular, it follows that Z_∞ is independent of the input and output impedance of this equivalent amplifier. Thus, when calculating Z_∞, we may choose these impedances arbitrarily for the sake of simplifying the analysis, let us say, as 0 or ∞.

In the system of Fig. 1.7, $\beta = k_2' \beta_0 k_2$, so, from (1.18), we have

$$\lim_{w_{43} \to \infty} K_\beta = -\frac{k_1}{k_2} \frac{k_1'}{k_2'} \frac{1}{\beta_0} + k_d \tag{1.20}$$

where $|k_d|$ is negligible for most problems of practical significance.

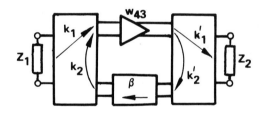

FIG. 1.7

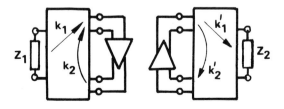

FIG. 1.8

The three multipliers of the first component of the (1.20) relate respectively to the input six-pole, output six-pole, and the four-pole β_0. In this way, the feedback amplifier design breaks into the three subsystem designs, which are nearly independent of each other:

The input circuit design must satisfy the specifications for the input impedance and noise figure (see also Sec. 1.5);

The output six-pole must ensure the specified output impedance and efficiency;

The β_0-circuit design implements the prescribed closed-loop transmission function.

When designing these circuits, certain trade-offs must be accepted between the target parameters and the available stability margins. The overall correction of the loop transmission function to provide the system stability can be accomplished, as previously mentioned, in the forward-path w_{43} without affecting the already designed input, output, and feedback circuits.

1.1.3 Inverse Block Diagram; Two-Poles' Connection

The combiner in Fig. 1.1 realizes the equation $E = U_1 - TE$. Hence, it follows that $U_1 = E + TE = (T+1)E = FE$ as diagramed in Fig. 1.9. Regardless of changing the direction of signal transmission, the formulas (1.1), (1.2), and (1.3) remain valid for the new system.

In the case of negative feedback, when $|T+1| > 1$, introducing the channel T increases U_1 and lowers the ratio E/U_1. Large $|K|$ makes the feedback large, $|T| \gg 1$, so that $U_1 \cong ET$ and $U_2/U_1 \cong -1/\beta$. Thus, all the features of the feedback equations (1.1), (1.2), and (1.3) are preserved.

Equations (1.1), (1.2), and (1.3) also serve to describe the interrelationship of the signals in the two-poles' connections depicted in Fig. 1.10. Table 1.1 indicates the definitions of the terms used.

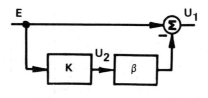

FIG. 1.9

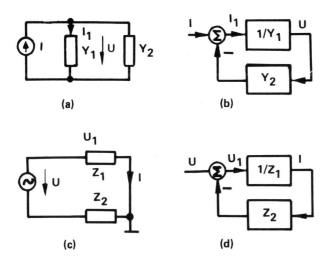

FIG. 1.10

TABLE 1.1

Fig. 1.1 (a)	Fig. 1.10 (a)	Fig. 1.10 (c)
U_1	I	U
U_2	U	I
E	$UY_1 = I_1$	$IZ_1 = U_1$
K	$1/Y_1$	$1/Z_1$
β	Y_2	Z_2
T	Y_2/Y_1	Z_2/Z_1
K_β	$1/(Y_1 Y_2)$	$1/(Z_1 Z_2)$

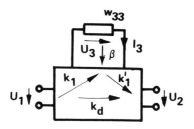

FIG. 1.11

The system of Fig. 1.11 is a modification of the system shown in Fig. 1.4, where the active element w_{33} is bilateral. Assume the following: $w_{33} = I_3/U_3$; k_1 is the voltage transfer coefficient from the system input to the open poles 3-3; k'_1 is the transfer impedance (of the output/input type) from the active element to the system output; k_d stands for the direct propagation coefficient under the condition of poles 3-3 open (i.e., $w_{33} = 0$); β represents the β-circuit driving-point impedance faced by the active element. The ouput can be viewed as the superposition of the effects caused by U_1 and the source $I_3 = U_1 k_1/(w_{33}^{-1} + \beta)$, which can replace the element w_{33} by virtue of the compensation theorem. The output/input ratio is given by

$$\frac{U_2}{U_1} = \frac{k_1}{1/w_{33} + \beta} k'_1 + k_d = \frac{k_1 k'_1}{\beta} \frac{w_{33}}{w_{33} + 1} + k_d =$$

$$= \frac{k_1 k'_1}{\beta} \frac{T}{T+1} + k_d \qquad (1.21)$$

where T stands for $w_{33}\beta$. If the feedback $|T+1|$ is large, the ratio U_2/U_1 does not depend on w_{33}. Formula (1.21) is, therefore, the analog of (1.18) for the case of bilateral element w_{33}.

The analogy follows from the general Bode expression for the return difference for the specified parameter of a selected circuit element (driving-point imittance, or a transfer function of a unilateral two-port) as $F = \Delta/\Delta°$, through the main determinant Δ of the circuit, and the minor $\Delta°$ to which Δ reduces if the parameter of the chosen element equals 0. The return ratio for the element is $T = F - 1 = \Delta/\Delta° - 1 = (\Delta - \Delta°)/\Delta°$. The specified parameter enters as a multiplier in the difference $\Delta - \Delta°$, and therefore in T.

1.1.4 Return Ratio as a Function of the Load

In this section, we recruit (1.47) from Sec. 1.4.2 to describe the dependence of a circuit function W on a parameter w. If we take W to be the return ratio T, and the variable parameter w to be the impedance of an external two-pole (load) Z_L, then

$$T(Z_L) = \frac{Z_0 T(0) + Z_L T(\infty)}{Z_0 + Z_L} \tag{1.22}$$

Replacing T with $F - 1$, and Z_0, with $ZF(\infty)/F(0)$, yields

$$F(Z_L) = \frac{Z + Z_L}{Z/F(0) + Z_L/F(\infty)} \tag{1.23}$$

It is instructive to list the particular forms these two equations accept with specific values of Z_L:

$$T(-Z_\infty) = 0 \tag{1.24}$$

$$T(-Z_0) = \infty \tag{1.25}$$

$$T(Z_\infty) = \frac{2 T(0) T(\infty)}{T(0) + T(\infty)} \tag{1.26}$$

$$T(Z_0) = \frac{T(0) + T(\infty)}{2} \tag{1.27}$$

$$F(Z) = \frac{F(0) F(\infty)}{F(0) + F(\infty)} \tag{1.28}$$

$$F(-Z) = 0 \tag{1.29}$$

$$F(Z_\infty) = \frac{2 T(0) T(\infty) + T(0) + T(\infty)}{T(0) + T(\infty)} \tag{1.30}$$

1.1.5 Balanced Bridge

As can be seen from Blackman's formula (1.16), each one of the following equations:

$$\left. \begin{array}{l} T(0) = T(\infty) \\ \beta(0) = \beta(\infty) \\ Z = Z_0 \end{array} \right\} \qquad (1.31)$$

presents a necessary and sufficient condition for Z to be invariant with respect to w_{43}. By virtue of (1.22), these are also necessary and sufficient conditions for β to be independent of the external two-pole impedance Z_L coupled to the terminal where Z had been measured. Therefore, the impedance of the input Z is invariant with respect to w_{43} if and only if T is invariant with respect to the source impedance Z_1. This set of conditions is known as *balanced bridge*.

Another necessary and sufficient condition for this case to exist is as follows: the transfer coefficient is 0 either from the output of the amplifier to the poles where Z is measured, or from these poles to the input of the amplifier. To prove this, characterize the β-circuit by the transmission coefficients k_1, k_3, and k_4 as indicated in Fig. 1.12. Calling upon the same logic used to construct (1.21), gives

$$\beta(Z_1) = k_3 k_4 \frac{1}{1/Z_1 + Z_0} + k_4$$

Then,

$$\beta(0) = k_4$$

$$\beta(\infty) = \frac{k_3 k_1}{Z_0} + k_4$$

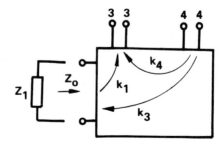

FIG. 1.12

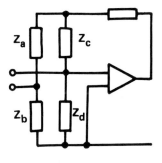

FIG. 1.13

and (1.31) yields $k_3 k_1 = 0$.

An example of a balanced bridge circuit is shown in Figure 1.13. The bridge is balanced if $k_3 = 0$, e.g.

$$Z_a / Z_b = Z_c / Z_d$$

Note that Z_∞ does not depend on Z_d as was proved previously, and therefore is expressed through Z_d, Z_b, and Z_c.

An effective method for calculation of Z_∞ is the imaginary balancing of the bridge by assigning a value to the input impedance of the amplifier such that $k_3 = 0$. Under this provision $Z_\infty = Z_0$, so Z_∞ can be calculated with the amplifier shut off. For the system of Fig. 1.13, $Z_d = Z_b Z_c / Z_a$ and

$$Z_\infty = Z_b \frac{Z_d + Z_b}{Z_d + Z_c}$$

The input impedance in a system with large feedback is $Z \cong Z_\infty$ and can be calculated as in the above. The sufficient value of the feedback can be estimated with (1.44) which can be found in Sec. 1.2.3.

For calculating the output impedance, zero transmission between the system's output terminals and the input of the amplifier is similarly useful. An example can be derived by reversing the direction of the amplifier in Fig. 1.13.

Figure 1.14 illustrates the use of directional couplers, each consisting of a three-winding (hybrid) transformer, or a pair of two-winding transformers (usable with 0.1-inch binocular magnetic cores up to 2 GHz), in feedback amplifiers. If the input (output) impedance of the amplifier element has the value required for the coupler to be balanced so that k_3 (or k_1) equals 0, or when the feedback is significant, w_{43} does not influence the input (or output) impedance of the system. It will be illustrated in the next section

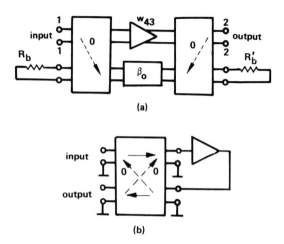

FIG. 1.14

that in the presence of large feedback, the system of Fig. 1.1.4(b) makes a so-called ideal power converter, also known as a transparent amplifier, which is a two-port having two specific features: the input impedance of the two-port equals the output load impedance and the two-port output impedance equals the internal impedance of the signal source.

1.1.6 Examples and Exercises

1. (a) What is the type of the feedback (positive, negative, or large) when $T = 40$; -40; $1 \angle 45°$; $1 \angle 225°$; 0.2; -0.2?
(b) Over what range of $|T|$ is the type of feedback not dependent on arg T?
(c) $T(s) = 10(5 + s)/[(0.5 + s)(2 + s)(4 + s)(20 + s)]$, where s is the operational variable; by substituting $s = j\omega$, calculate the frequency regions of positive, negative, and large feedback.

2. Find the transmission function of the link depicted in Fig. 1.15(a).

3. Symmetrical Π-pulses are applied to the input of the feedback system of Fig. 1.15(b), the main Fourier components of which belong to the frequency range of large feedback. What are the spectrum components and the shape of the signals at the output and the input of the lowpass filter Q?

4. Fig. 1.15(c) shows a system with pure time-delay τ in the feedback loop. Assuming the feedback to be large, characterize the signal at the input of the delay line. Is it possible to physically realize such a system? What kinds of approximations are feasible?

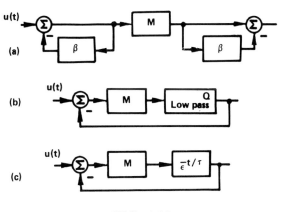

FIG. 1.15

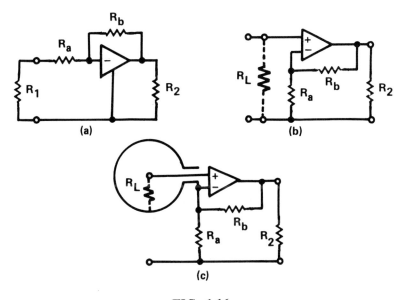

FIG. 1.16

5. In the feedback systems of Fig. 1.16 (a, b, c), the amplifier possesses very large input impedance, very low output impedance, and voltage transfer coefficient equal to μ. R_l stands for the surface leakage resistance.

Assuming $R_l = \infty$, find the expressions for $T(0)$, $T(\infty)$, T, Z_0, and Z relating to the input terminals for the system of Fig. 1.16(a, b). How does the value of R_l affect the input impedance? Why does application of a conducting ring, as shown in Fig. 1.15(c), increase the input impedance?

6. Determine F for the system shown in Fig. 1.17(a).

For the sake of convenience, redraw the schematic as Fig. 1.17(b) shows. Denoting the input impedance of the transistor as h_{11}, and neglecting the impedance Z_c as being small compared to the output impedance of the transistor, the current transfer coefficient of the remining L-network is

$$\frac{Z_e}{Z_e + Z_b + h_{11}}$$

Denoting the gain in current of the transistor as $-h_{21}$, the return difference is given by

$$F = T + 1 = h_{21} \frac{Z_e}{Z_e + Z_b + h_{11}} + 1 \qquad (1.32)$$

7. Find the expression for the input impedance Z of the circuit presented in Fig. 1.17(c).

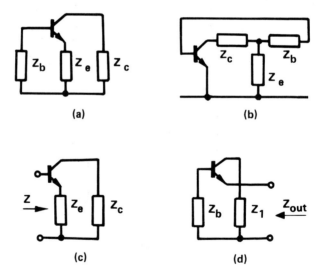

FIG. 1.17

By making $h_{21} = 0$, i.e., by disconnecting the collector terminal, we find that $Z_0 = h_{11} + Z_e$. Further, allowing Z_b to be infinite, we find from (1.32) that $F(\infty) = 1$; letting Z_b equal 0,

$$F(0) = \frac{h_{21} Z_e}{Z_e + h_{11}} + 1$$

Finally, using Blackman's formula,

$$Z = h_{11} + Z_e(h_{21} + 1)$$

8. Find the output impedance Z_{out} for the system of Fig. 1.17(d).

The output impedance without the feedback (with the collector terminal disconnected) is $h_{11} + Z_b$. Relative to the input terminal $F(0) = 1$, and $F(\infty)$ is equal to the previously obtained value of F in (1.32). Next, applying (1.16) gives

$$Z_{out} = \frac{h_{11} + Z_b}{h_{21} + 1}$$

9. Express the output impedance of a common-drain stage employing a field-effect transistor with transconductance $y_{21} = h_{21}/h_{11}$. (Use the previous results for a bipolar transistor.)

10. Find the input impedance Z and the impedance Z_{cc} between disconnected poles c-c of the feedback amplifier of the schematic given in Fig. 1.18, where $h_{21} = 50$ and $h_{11} = 1$ kΩ for all the transistors, and $R_1 = R_2 = R_c = 2$ kΩ, $R_a = R_d = 100$ kΩ, $R = 10$ kΩ.

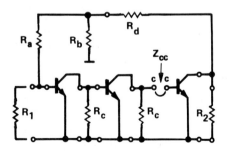

FIG. 1.18

The approximate transfer ratio can be readily expressed using the current transfer coefficients of the elementary links:

$$T(R_1) = T \cong h_{21}^3 \left(\frac{R_c}{R_c + h_{11}} \right)^2 \frac{R_2}{R_d} \frac{R_b}{R_b + R_a} \frac{R_1}{R_1 + R_{11}} = 70$$

From this, $T(0) = 0$, and

$$T(\infty) \cong h_{21}^3 \left(\frac{R_c}{R_c + h_{11}} \right)^2 \frac{R_2}{R_2 + R_d} \frac{R_b}{R_b + R_d} = 100$$

Without the feedback, the input impedance is $Z_0 = h_{11}$. Then, from (1.16), $Z \cong 10\,\Omega$.

With respect to the poles $c\text{-}c$, $Z_0 = R_c + h_{11}$, $T(\infty) = 0$, and $T(0)$ is equal to the value denoted before as $T(R_1)$. Finally, using (1.16), $Z_{cc} \cong 200$ kΩ.

11. Using current transfer coefficients of elementary links, calculate the return ratios and the input and output impedances in the amplifiers with emitter feedback shown in Fig. 1.1(a).

12. Consider, crudely or quantitatively, the dependence of T on the impedances of the sources and loads in the system of Figure 1.19. What is the effect of cascading the systems on the feedback in each of them? Which tandem connections are most practical?

13. In the system of Fig. 1.20, determine the kind of feedback relative to the poles 1-1, and $a\text{-}a$. Calculate the input impedance in two different ways:
   a. directly, using Blackman's formula;
   b. by adding Z_a to the impedance calculated for the remainder of the input contour.
Conduct a similar analysis for the system of Fig. 1.20(b).

14. Using Blackman's formula or the method of imaginary balancing of the bridge, show that for the input of the system of Fig. 1.20(c), $Z_\infty = Z_a Z_b / Z_2$. Consider the frequency responses of this impedance related to various types of Z_a, Z_b, and Z_2.

15. Using (1.16) or the method of imaginary bridge balancing, calculate the input impedance Z_∞ of the feedback amplifier display in Fig. 1.21, which serves as a load for an LC filter-corrector for a signal source having capacitive internal impedance. The feedback is assumed large.

16. Using (1.16) or the method of imagined bridge balancing, show that the input impedance of the system with large feedback depicted in Fig. 1.22(a) is given by $Z_\infty = Z_b / mn$, and for the system of Fig. 1.22(b) by

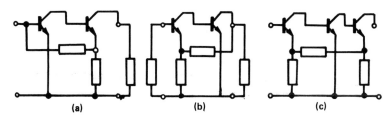

FIG. 1.19

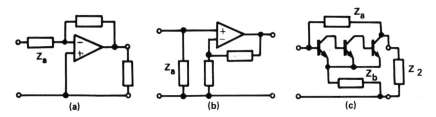

FIG. 1.20

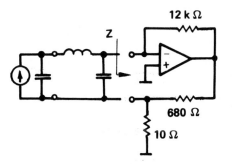

FIG. 1.21

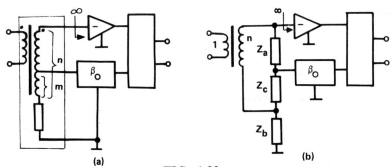

FIG. 1.22

$$Z_\infty = \frac{Z_b(Z_a + Z_c)}{n^2(Z_c + Z_b)}$$

Compare Fig. 1.22 with Fig. 1.14(a).

17. Find the dependence of T on the load impedance of a voltage-follower built around a field-effect transistor.

18. Input impedance of an amplifier is $Z_0 = 1\,\text{k}\Omega\,\underline{/-45°}$, $T(0) = 200\,\text{k}\Omega\,\underline{/\,160°}$, $T(\infty) = 0$; internal impedance of the signal source is $32\,\text{k}\Omega\,\underline{/\,-90°}$ (for a capacitance of 0.05 pF at 100 MHz). Find the feedback.

19. In an amplifier whose input had been matched with the source impedance Z_0, series feedback was introduced, which reduced the input signal $|F(Z_0)|$ times and increased the input impedance $|F(0)|$ times, thus eliminating the matching and causing input mismatch losses. Decide whether it is advantageous to restore the matching by application of a transformer with transformation ratio $|F(0)|^{1/2}$ to 1.

Using (1.23) with $Z_L = Z_0$, $F(\infty) = 1$, $Z = Z_0 F(0)$, the return difference in the system without a transformer is $F(Z_0) = [F(0) + 1]/2 = F(0)/2$. After the transformer was introduced, Z_L accepts the value $Z_0 F(0)$ and the return difference becomes $F[Z_0 F(0)] = 2/[1/F(0) + 1] \cong 2$. Thus, the transformer application causes the feedback to drop $|F(0)|/4$ times. At the same time, the input voltage increases, but only $2|F(0) + 1|/|F(0)|$ times. Therefore, introducing the transformer proves to be a disadvantage.

20. A two-port is said to be a negative impedance converter (NIC) if its input impedance repeats the impedance of the load, with opposite sign, and either the input voltage equals the output voltage, or the input current equals the output current.

Show that in a NIC realized as a feedback system, the return ratio $T = -1$ when the loads connected to the input and output terminals are equal.

Show that if the gain of the amplifier element in a feedback system is sufficient, and connecting equal termination impedances to the input and the output of the system entails a small value of T, then, this system's input (output) impedance equals, with opposite sign, the load (signal source) impedance.

Show that with large gain amplifiers any of the systems in Fig. 1.23 (a, b, c) will realize a NIC.

21. The ideal power converter (IPC) is a two-port with K-times voltage and current amplification, and the input impedance equal to the load impedance as well as the output impedance equal to the source internal impedance, as indicated in Fig. 1.24. Show that the circuit presented in Fig. 1.25, utilizing a noninverting amplifier with infinite input and output impedances and the transimpedance $(K-1)/Z$, emulates an IPC [123].

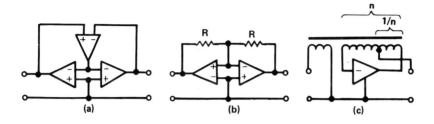

FIG. 1.23

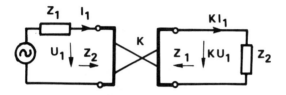

FIG. 1.24

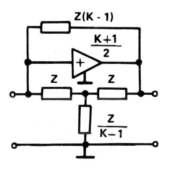

FIG. 1.25

22. Show that if the amplifier gain is large, the systems of Fig. 126 [98,103] represent realizations of an IPC (recall *Prob. 16*, setting $mn = 1$). What difference can be seen between the systems of Fig. 1.26 and Fig. 1.23(c)?

23. Describe the features of a mechanical manipulator that consists of a cascade connection of an IPC and a transformer, with electromechanical transducers at both sides.

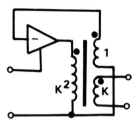

FIG. 1.26

1.2 SENSITIVITY

1.2.1 Sensitivity of Transmission Function

The *sensitivity* of closed-loop transfer coefficient to w_{43} is given by

$$S = \frac{d K_{E\beta}/K_{E\beta}}{d w_{43}/w_{43}} = \frac{d \ln K_{E\beta}}{d \ln w_{43}} = \frac{d\, 20 \log K_{E\beta}}{d\, 20 \log w_{43}} \qquad (1.33)$$

This widely accepted definition is the inverse of that originally proposed by Bode.

The sensitivities for K_β and $K_{I\beta}$ correspondingly involve $F(0)$ and $F(\infty)$ in place of F.

Denoting $S = S_m + jS_\phi$, where S_m and S_ϕ are real, we rewrite (1.33) as

$$d \ln |K_{E\beta}| + jd \arg K_{E\beta} = (S_m + jS_\phi)(d \ln |w_{43}| + jd \arg w_{43})$$

so that

$$d \ln |K_{E\beta}| = S_m d \ln |w_{43}| - S_\phi d \arg w_{43} \qquad (1.34)$$

$$d \arg K_{E\beta} = S_\phi d \ln |w_{43}| - S_m d \arg w_{43} \qquad (1.35)$$

Substituting (1.10) into (1.33) yields

$$S = \frac{\dfrac{K_{E\beta} - k_{Ed}}{K_{E\beta}} \dfrac{d K_E}{K_{E\beta} - k_{Ed}}}{d w_{43}/w_{43}} = \left(1 - \frac{k_{Ed}}{K_{E\beta}}\right) \frac{1}{F}$$

Most often, the direct propagation coefficient k_{Ed} is negligible, and the sensitivity is, therefore, the inverse of the return difference:

$$S \cong 1/F \tag{1.36}$$

$$S_m \cong \operatorname{Re} \frac{1}{F} = \frac{1 + |T| \cos \phi}{1 + 2|T| \cos \phi + |T|^2} \tag{1.37}$$

$$S_\phi \cong \operatorname{Im} \frac{1}{F} = \frac{-|T| \sin}{1 + 2|T| \cos \phi + |T|^2} \tag{1.38}$$

where $\phi = \arg T$.

Figure 1.27 shows $|TS_m|$ and $|TS_\phi|$ plotted *versus* $|\phi|$. Using these diagrams, S_m and S_ϕ can be found from known values of $|T|$ and $|\phi|$.

The first component in (1.34) and (1.35) is usually dominant within the band of operation so that S_m characterizes deviations from the nominal system gain, and S_ϕ from the system's phase shift. It can be seen from Fig. 1.7 that S_m attains its minimum when $\phi = \pi/2 + \arcsin 1/|T| \cong \pi/2$, and S_ϕ, when $\phi = 0$ or $|\phi| = \pi$. An increase in $|T|$ reduces the deviation in both the gain and phase shift.

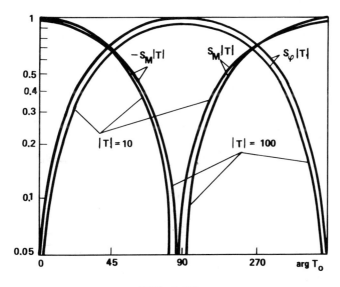

FIG. 1.27

Strictly speaking, the tolerances of the closed-loop transfer function may be evaluated using the sensitivity when the variations of w_{43} are small (let us say, less than 5 dB in changes of $|w_{43}|$). However, the widely employed methods of feedback systems design aimed at sensitivity minimization, i.e., at feedback maximization, remain practical when the tolerances in w_{43} are much larger, since, as shown by (1.6) and (1.7), large feedback provides for system precision operation in this case as well.

1.2.1 Sensitivity of Driving-Point Impedance

By derivating (1.6) and substituting $T(0) = w_{43}\beta(0)$ and $T(\infty) = w_{43}\beta(\infty)$, we obtain

$$dZ = d\left(Z_0 \frac{1 + w_{43}\beta(0)}{1 + w_{43}\beta(\infty)}\right) = Z_0 \frac{\beta(0) - \beta(\infty)}{[1 + w_{43}\beta(\infty)]} dw_{43}$$

or

$$dZ = Z_0 \frac{F(0) - F(\infty)}{[F(\infty)]^2} \frac{dw_{43}}{w_{43}}$$

Dividing this equation into (1.6) gives

$$\frac{dZ}{Z} = \frac{F(0) - F(\infty)}{F(0) F(\infty)} \frac{dw_{43}}{w_{43}}$$

from which the *sensitivity of the driving-point impedance* is

$$S_z = \frac{dZ/Z}{dw_{43}/w_{43}} = \frac{F(0) - F(\infty)}{F(0) F(\infty)} \quad (1.39)$$

IF both $|F(0)| \gg 1$ and $|F(\infty)| \gg 1$, the sensitivity

$$S_z = \frac{1}{w_{43}} \frac{\beta(0) - \beta(\infty)}{\beta(0) \beta(\infty)} \quad (1.40)$$

As $F(0)$ approaches $F(\infty)$, the driving-point impedance sensitivity decreases. If $F(0) = F(\infty)$, then $S_z = 0$, which reflects the feature of the balanced bridge structure.

1.2.3 Reflection Coefficients

Using (1.16), the reflection coefficient between Z and Z_0 is given by

$$\frac{Z - Z_0}{Z + Z_0} = \frac{F(0) - F(\infty)}{F(0) + F(\infty)} \tag{1.41}$$

between Z_∞ and Z_0, by

$$\frac{Z_\infty - Z_0}{Z_\infty + Z_0} = \frac{T(0) - T(\infty)}{T(0) + T(\infty)} \tag{1.42}$$

and between Z_∞ and Z, which can be straightforwardly calculated, by

$$\frac{Z_\infty - Z}{Z_\infty + Z} = \frac{T(0) - T(\infty)}{2\,T(0)\,T(\infty) + T(0) + T(\infty)} \tag{1.43}$$

Combining (1.30), (1.41), and (1.43) we arrive at the equation:

$$\frac{Z_\infty - Z}{Z_\infty + Z} = \frac{1}{F(Z_\infty)} \frac{Z_\infty - Z_0}{Z_\infty + Z_0} \tag{1.44}$$

indicating that the deviations of the input impedance Z from the value of Z_∞ (which is related to the case of infinite feedback) are inversely proportional to the value of the feedback calculated while the external load impedance is Z_∞.

When the input combiner is designed using the simplified formula (1.17) and the impedance Z_∞ is matched to the source impedance Z_1, the input reflection coefficient is equal to this coefficient, calculated when no feedback is applied, divided by the feedback. The bigger the feedback, then the better the matching that is attained.

From (1.44), it follows that the equalities:

$$Z = Z_0$$

and

$$F(Z_\infty) = 1 \tag{1.45}$$

are equivalent, thus supplying an extra condition for the input bridge to be balanced, supplementary to (1.31).

1.2.4 Examples and Exercises

1. In a feedback system, the return ratio is $T = 30 \,\underline{/\text{-}150°}$.
(a) Determine S, S_m, S_ϕ (use the diagrams of Fig. 1.24).
(b) What is the change in K_β if $|w_{43}|$ drops by 3 dB (use the approximate approach with S_m)?
(c) What increment in K_β results from $|w_{43}|$ increasing indefinitely (use Eq. (1.4)?
(d) What increment in K_β results from $|w_{43}|$ increasing such that T becomes $100 \,\underline{/\text{-}150°}$ (use Eq. (1.7))?

2. The forward path in the feedback system of Fig. 1.1(a) is composed of the link M_1 followed by two parallel links, M_2 and M_3. Find the sensitivities of the closed-loop transfer function to M_1, M_2, and M_3. Compare the sensitivities while suggesting various phase shifts in these links.

3. Write down the expression for the sensitivity of the current transfer coefficient $K_{I\beta}$.

4. Determine the change in percent in the input resistance of a feedback system reflecting a 10 percent change in w_{43}, if $F(0) = 30$ and $F(\infty) = 50$ (all values are assumed to be real).

5. The resistance of an external load of 50 Ω produces a reflection coefficient of 0.6 when the feedback path is disconnected. The system is designed such that $Z_\infty = 50$ Ω. Determine the minimal value of the feedback sufficient for the reflection coefficient not to exceed 0.05.

6. Extend the definition of formulas (1.41) through (1.44) to the case of a two-pole element w_{33} (instead of two-port w_{43}), using the analogy described in Sec. 1.1.3.

1.3 NONLINEAR DISTORTION

The output of the nonlinear forward path in response to the sinusoidal input with the frequency ω comprises the fundamental with the amplitude I_4 and additional Fourier components, called nonlinear products. Let us consider one of these products having the frequency ω_n and the amplitude I_n. The ratio in amplitudes of the nonlinear product to the fundamental at the output of the system with a cross-sectional feedback path is

$$q = \left| \frac{I_n \, k'_{1n}}{I_4 \, k'_1} \right|$$

the subscript n indicating the frequency ω_n.

If $q \ll 1$, we may consider the forward path to be approximately linear, with the source of current I_n connected to its output in parallel. The amplitude of the nonlinear product at the system's output in the presence of the feedback:

$$U_{n2} = \left| \frac{k'_{1n} I_n}{F_n} \right| = \left| \frac{q}{F_n} k'_1 I_4 \right| = \left| \frac{q}{F_n} \frac{k_1 k'_1 w_{43}}{F} E \right| = \left| \frac{q}{F_n} (K_{E\beta} - k_{Ed}) \right| E$$

is reduced by the feedback occurring at the frequency of the product. The nonlinearity coefficient of the closed-loop system is, therefore, given by

$$q_\beta = \left| \frac{U_{n2}}{U_2} \right| = \left| \frac{q}{F_n} \left(1 - \frac{k_d}{K_{E\beta}} \right) \right| \tag{1.46}$$

If the system's parameters are the same at the frequencies ω and ω_n, then $q_\beta = |S| q$.

1.4 REGULATION

1.4.1 Introduction

The plant characteristics may be largely dependent on a variable parameter, such as the cable temperature in telecommunication systems, or the payload thermal inertia in a temperature-controlled furnace, *et cetera*.

Correspondingly, feedback system performance can be enhanced by adjusting (tuning) the transfer function of the compensator in Fig. 1.1(b) in response to the variations of the plant parameters, measured or calculated in advance. For this reason, we will consider below the theory of the Bode regulator (variable equalizer) [21].

It is also of interest that such a regulator, where the variable linear element has been replaced by a nonlinear nondynamic element, forms a dynamic nonlinear link with a very broad range of behavior. This circuit serves, in particular, as a point of departure in the iterative synthesis procedure of a nonlinear dynamic compensator (NDC) for a feedback loop (presented in Chapter 5).

1.4.2 Parameter Dependence of a Circuit Function

Let W represent either a driving-point immittance (i.e., impedance or admittance [23]), or a transfer immittance, or a transfer coefficient of current or voltage, whereby all of these functions are described as a ratio of the incident signal to the response, or the inverse. Let w stand either for the immittance of variable two-pole, or for the variable transfer immittance (coefficient) of unilateral amplifier. Under these assumptions the function $W(w)$ is bilinear [23], mapping any circumference (or a straight line) from w-plane onto a circumference (or a straight line) in the W-plane. This function may be expressed as

$$W = \frac{w_1 W(0) + w W(\infty)}{w_1 + w} \tag{1.47}$$

or as

$$W + \frac{w_2 + w}{w_2/W(0) + w/W(\infty)} \tag{1.48}$$

where $W(0)$ stands for W when $w = 0$, and $W(\infty)$, for W when $w = \infty$. The parameters w_1 and w_2 can be found from appropriate boundary conditions; for example, the equality:

$$W + \infty \tag{1.49a}$$

implies that

$$w_1 + w = 0 \tag{1.49b}$$

and the equality:

$$W = 0 \tag{1.50a}$$

suggests that

$$w_2 + w = \infty \tag{1.50b}$$

Provided that W stands for a ratio of the type of signal to response, the condition (1.50a) indicates that the circuit faces oscillation. Consequently, each contour impedance is 0 and the return ratio for each of the amplifiers in this circuit equals -1. Therefore, it follows from (1.50b) that if w describes the impedance of a two-pole, the parameter w_2 represents the driving-point impedance between the terminals to which the two-pole w is connected. If w designates the transfer coefficient of an amplifier, then w_2 must equal $-1/\beta$, where β represents the feedback-path transmission coefficient for this amplifier.

In a similar way, the coefficients w_1 and w_2 can be determined for any conceivable meaning of W and w, thereby generating, in particular, many of well known formulas, such as (1.10), (1.16), (1.21), *et cetera*.

Using this kind of reasoning, or exploiting (1.17) and the analogy discussed in Sec. 1.1.3, we readily arrive at the relation:

$$\frac{W_1(0)}{W_1(\infty)} = \frac{W_2(0)}{W_2(\infty)} \tag{1.51}$$

which is sustained for the circuits shown in Fig. 1.28. Here, $W_1(0)$ denotes the value of $W_1(w_1)$ determined while connecting a two-pole with zero driving-point immittance (or an amplifier with zero transfer coefficient or immittance) to the opposite side of the network. The values $W_1(\infty)$, $W_2(0)$, and $W_2(\infty)$ are defined in a similar way.

The sensitivity can be expressed either using (1.47) as

$$S = \frac{d \ln W}{d \ln w} = \frac{w_1 w [W(\infty) - W(0)]}{(w_1 + w)[w_1 W(0) + w W(\infty)]} \tag{1.52}$$

or, using (1.48), as

$$S = \frac{w[w_2/W(0) - w/W(\infty)]}{(w_2 + w)[w_2/W(0) + w/W(\infty)]} \tag{1.53}$$

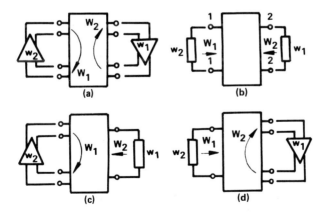

FIG. 1.28

1.4.3 Symmetrical Regulation

Regulation is called symmetrical with respect to a certain nominal value w_0 of the variable parameter w when the maximum relative deflections of $W(w)$ from $W(w_0)$, up and down, are equal, i.e., when the function of regulation:

$$Q = \frac{W(\infty)}{W(w_0)} = \frac{W(w_0)}{W(0)} \qquad (1.54)$$

By substituting this expression into (1.47), we have

$$w_0 = w_1 / Q$$

and

$$W = W(w_0) \frac{1 + (w/w_0) Q}{(w/w_0) + Q} \qquad (1.55)$$

which is illustrated in Fig. 1.29. Regulators with such characteristics are called Bode variable equalizers [22]. The sensitivity of W to w in these regulators possesses an extremum at $w = w_0$, as can be seen by substituting (1.54) into (1.52).

Denoting

$$\Gamma = \frac{w - w_0}{w + w_0} \frac{Q - 1}{Q + 1} = \frac{W - 1}{W + 1} \qquad (1.56)$$

the function W is expressed as

$$W = W(w_0) \frac{1 + \Gamma}{1 - \Gamma} \qquad (1.57)$$

and

$$\ln W = \ln W(w_0) + 2 \text{ Arc th } \Gamma \qquad (1.58)$$

The Taylor expansion of the preceding expression is

$$\ln W = \ln W(w_0) + 2 \left(\Gamma + \frac{\Gamma^3}{3} + \frac{\Gamma^5}{5} + \dots \right) \qquad (1.59)$$

The expansion truncated after the first term gives a very accurate approximation to ln W up to fairly large values of $|\Gamma|$. For example, with $|\Gamma| = 1/2$, the regulation constitutes 1 Np (8.7 dB), and the maximal error due to omitting the second term of the series is only $(2/3)/2 = 0.083$ Np (0.72 dB).

The variable corrector comprising an auxiliary four-pole is depicted in Fig. 1.30. Denoting the image parameters [60] of the four-pole as $g_c = a_c + jb_c$, z_{c1}, and z_{c2}, the input impedance is given by

$$W = z_{c1} \frac{1 + (w/z_{c2}) \coth g_c}{w/z_{c2} + \coth g_c} \quad (1.60)$$

Equations (1.55) and (1.60) coincide if $w_{c1} = W(w_0)$, $w_{c2} = w_0$, and $g_c =$ Arc coth Q. In this case, $(Q-1)/(Q+1) = \exp(-2g_c)$, and Γ conforms to the

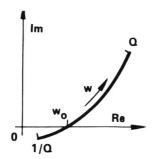

FIG. 1.29

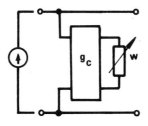

FIG. 1.30

definition of reflection coefficient between w and w_c. The regulation in $|W|$ is then given by

$$\ln\left|\frac{W(w)}{W(w_0)}\right| = 2\operatorname{Re}\Gamma = 2\,\frac{w - w_0}{w + w_0}\exp(-2a_c)\cos 2b_c \qquad (1.61)$$

Most frequently, w is chosen to be real, and is realized by a pin-diode, field-effect transistor, thermistor, *et cetera*.

After selecting an appropriate structure for the auxiliary four-pole (usually, L-, T-, or Π-type), the impedances of the branches may be found by successive approximation.

The simplest corrector with the L-type auxiliary four-pole shown in Fig. 1.31 can be designed in a straighforward manner. If w changes from 0 to ∞, and $w_0 = 1$ (thus, we normalize the impedances relative to the intermediate impedance of the variable element), then

$$Z_1 = Q - 1/Q \qquad (1.62)$$

$$Z_2 = 1/Q \qquad (1.63)$$

The realizability conditions for this corrector are that Z_1 and Z_2 are positive real functions (p.r.f.) of s, i.e. $\operatorname{Re} Z_1(j\omega) > 0$, $\operatorname{Re} Z_2(j\omega) > 0$, which are conditions equivalent to

$$\begin{aligned}|\arg Q(j\omega)| &< \pi/2 \\ |Q| &> 1\end{aligned} \qquad (1.64)$$

A similar theory can be developed for a regulator employing a variable gain amplifier as the variable element.

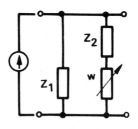

FIG. 1.31

1.5 NOISE

1.5.1 Noise at the System's Output

In the feedback system of Fig. 1.32, N_1 represents the total noise of the amplifier μ and the feedback path (including the sensor noise), and N_2 is the noise generated inside the plant.

The system's total output is given by

$$(k_1 U_1 + N_1 + N_2/\mu)\, \frac{1}{\beta}\, \frac{T_0}{F_0} \tag{1.65}$$

We can see that increasing $|k_1|$ and $|\mu|$ magnifies the signal-to-noise ratio. The constraints on the values of $|k_1|$ and $|\mu|$ will be derived in the next chapter.

For the design of feedback systems with reasonably low output noise and maximized $|\mu|$ (maximized feedback), an iterative strategy is appropriate. The process of such a design rapidly converges if the subsequent design steps are in agreement with the initial data for the preceding steps which have already been executed, as in the following sequence:

Design the input circuit having the desired frequency response of k_1 in the working band, and sufficiently large $|k_2|$ at the higher frequencies beyond the working band (to provide the system stability; see also Chapter 2).

Design a link β/k_1 such that in the working frequency band the closed-loop transfer function k_1/β meets the system specifications, and the value of β/k_2 is sufficiently large at higher frequencies beyond the working band (see also Chapter 2).

Design a link μ so as to achieve maximized $|\mu|$ in the working band under the limitation imposed by stability conditions (to be studied in the next chapter), keeping the noise at the input of the nonlinear link bounded (to be discussed in the next section).

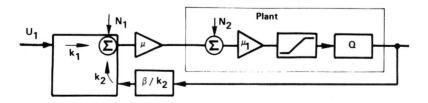

FIG. 1.32

1.5.2 Noise at the Input of the Plant

The noise at the input to the nonlinear link of the plant is of fundamental importance when a control system of the type shown in Fig. 1.32 is considered. The link Q represents the dynamics of the object of regulation with lowpass filter properties. The feedback signal is taken from a sensor that measures the position of an inertial object driven by an actuator of limited power, i.e., having the nonlinear characteristic of saturation.

Suppose the noise N_1 is Gaussian. Its effect on the input of the nonlinear link is given by

$$\frac{\mu_1 \mu_2}{F} = \begin{cases} |1/Q\beta)| N_1 & : |T| \gg 1 \text{ (lower frequencies)} \\ |\mu \mu_1| N_1 & : |T| \ll 1 \text{ (higher frequencies)} \end{cases} \quad (1.66)$$

The lower frequency noise components are independent of $\mu \mu_1$, and relatively small. The higher frequency noise components are proportional to $|\mu \mu_1|$. If $|\mu \mu_1|$ is excessively large, these components overload the nonlinear link (saturation link), thereby interfering with the actuator's signal transmission capability (reducing the output signal amplitude, causing excessive wear and tear in mechanical devices, *et cetera*)—notwithstanding that these noise components can be barely observed at the ouput of the lowpass Q (recall the comment in the Introduction).

For this reason, $|\mu \mu_1|$ must be bounded at higher frequencies. This limitation, in turn, constrains the feedback available in the working band, as will be shown in the next chapter.

1.5.3 Signal-to-Noise Ratio

Let k_{n1} and k_{nd} designate the transfer coefficients from the terminals that the source of noise N is applied to, as indicated in Fig. 1.33. Using (1.7), the ratio of the effects of the signal and the noise at the system's output is given by

$$\frac{U_2}{U_{n2}} = \frac{k_1 + k_d F/k_1'}{k_{n1} + k_{nd} F/k_1'} \frac{U_1}{N} \quad (1.67)$$

If the coefficients of direct propagation are negligible, as in large gain feedback amplifiers, the resulting ratio:

$$\frac{U_2}{U_{n2}} = \frac{k_1}{k_{n1}} \frac{U_1}{N} \quad (1.68)$$

does not depend on the feedback. In particular, this ratio does not change while breaking off the feedback path (the loading for the disconnected parts of the feedback circuit must be preserved, recall Sec. 1.1.2). Then, it is possible, and most often convenient, to examine the signal-to-noise ratio in the feedback system as if there were no feedback.

The signal-to-noise ratio is maximal if the signal source is connected to the amplifier input by a lossless network having the output impedance (as faced by the amplifier) equal to the value of Z_{opt} [156], which is specified as optimal for the amplifier element used. Preserving the matching at the system's input ($Z_1 = Z$) is generally not required for optimization of the signal-to-noise ratio.

If, in the system of Fig. 1.34, the impedance Z_{opt} faced by the input of the amplifier and the coefficient k_1 remain unvaried, then the signal-to-noise

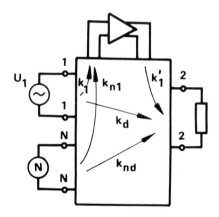

FIG. 1.33

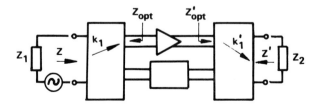

FIG. 1.34

ratio at the system's output remains unchanged, regardless of the type of feedback at the input terminals (series, parallel, or compound) and the resulting input impedance Z. Using compound feedback, the two separate problems of signal-to-noise ratio optimization and input to the source of signal matching (noise matching and power matching) can be solved simultaneously [122].

It is also worth mentioning that the design of the input six-pole is in many respects similar to the design of the output six-pole, where it is required that the amplifier's terminals again face a certain prescribed impedance, denoted Z'_{opt}, in order to attain maximum efficiency.

1.5.4 Examples and Exercises

1. The feedback in the multistage amplifier of Fig. 1.35 can be eliminated by shorting the resistor R_a, or by disconnecting the resistor R_b. In the first case, the output signal-to-noise ratio improves, in the second it decreases. Do these facts contradict the rule proved in Sec. 1.5.3 that the feedback, generally, does not influence the signal-to-noise ratio? What are the conditions of breaking the feedback path for this rule to be valid? Are the conditions satisfied in these examples?

2. Figure 1.36 shows the capacitive signal source, such as a photodiode, connected to a transimpedance amplifier. Parallel feedback reduces the input impedance of the amplifier to a value much smaller than the source impedance. Thus, while calculating the amplitude of the output signal, we may neglect the source capacitance, and find the output voltage as the product of the source current and the resistance of the feedback path resistor. However, the feedback does not improve (or make flat) the frequency response of signal-to-noise ratio, which is affected by the source capacitance and decays with frequency.

3. Assume that resistor R_1 is introduced in the input contour for the purpose of matching to the signal source having the same internal resistance. Given that equivalent noise temperatures of the source and the added resistor are considered equal, what is the resulting penalty in the maximum achievable signal-to-noise ratio?

To calculate the penalty, we introduce a transformer with a winding ratio of $1/\sqrt{2}$, converting the combination of the source with the added resistor back into the source with the same resistance R_1, but with $\sqrt{2}$ times (3 dB) smaller e.m.f. Hence, the penalty is 3 dB, i.e., the method of matching illustrated in Fig. 1.16 increases the minimum realizable noise figure by 3 dB.

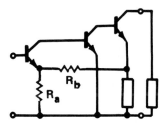

FIG. 1.35

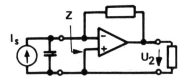

FIG. 1.36

Similarly, using the shunting resistor at the amplifier's input reduces the signal-to-noise ratio by 3 dB. The input matching can be secured without increasing the noise figure by employing compound feedback with small attenuation for the incident signal in the direction of k_1.

4. As shown in Sec. 1.1.6, compound feedback in the system with the three-winding transformer displayed in Fig. 1.22 guarantees the desired value of input impedance without an auxiliary series or shunt resistor. If the feedback winding ratio m is sufficiently small (less than $0.2\,n$), the penalty in the noise figure due to dissipation of the input signal in the resistor R_b is negligible. The compound feedback, therefore, provides a noise figure that is better by 3 dB than those with series or parallel auxiliary resistors.

5. Consider the feedback amplifier shown in Fig. 1.20(c), which has input impedance in the case of infinite feedback given by $Z_\infty = Z_a Z_b / Z_2$. Assuming all the values to be real and $R_a \gg R_2$, the equivalent circuit for the signal transmission to the input of the amplifier reduces to the attenuator depicted in Fig. 1.37, and the noise figure increases by the amount of attenuation for this attenuator, which is

$$\sim 20 \log \left(\frac{R_b}{2 R_{opt}} + \frac{R_{opt}}{2 R_a} \right)$$

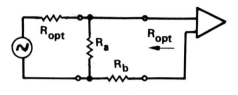

FIG. 1.37

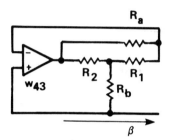

FIG. 1.38

In order to keep this increase small, R_b and $1/R_a$ must also be small, relative to R_{opt}.

The circuit of Fig. 1.20(c) is redrawn in Fig. 1.38 to show the structure of the β-circuit. Since R_b and $1/R_a$ are small, as noted before, then $\beta(0)$ and $\beta(\infty)$ are small as well. Therefore, a large amplifier gain is required so that $|F(0)| \gg 1$, $|F(\infty)| \gg 1$, and, consequently, $Z \cong Z_\infty$. This implies using multistage amplifiers and encountering certain difficulties in the provision of stability for wideband systems.

Create and consider a numerical example.

6. Direct propagation of the signal through a feedback circuit, when in phase with the output signal of the amplifier, helps reduce the noise figure in feedback amplifiers. Such an effect is noticeable in low-gain, single-stage feedback amplifiers with a hybrid transformer type of coupler in the feedback path [149].

7. In an ideal power converter (IPC, recall Sec. 1.1.4) having the gain coefficient K, the input port emulates a resistor with the equivalent noise temperature T/K when a resistor with the equivalent noise temperature T is connected to the IPC output. Which of the realizations of the IPC described in this chapter are suitable for this purpose? What factors limit the lowest available noise temperature of such a two-pole?

8. Resistor R at the temperature T connects the output and input of the noiseless amplifier with voltage gain $-K$, and high input and low output impedances. Show that the input impedance of the system is $R/(1+K)$, and its equivalent noise temperature is $T/(1+K)$.

9. A resistor at low temperature can be imitated by the feedback system of Fig. 1.39(a), comprising a low-noise amplifier with high input and output impedances. As can be found by using (1.16), the system input resistance is $R = R_b(1 + 1/m)$. Using the superposition principle, let us initially find the noise sources in the Thevinen equivalent two-pole shown in Fig. 1.39(b).

Consider first the mean square thermal noise e.m.f. $\overline{E}_{n1} = (4kT\Delta f R_b)^{1/2}$ of the resistor R_b. Here, T stands for the absolute temperature; k stands for the Boltzman constant; and Δf stands for the bandwidth. Assuming that the feedback is large, the effect of this noise at the amplifier input is small. As a consequence, the voltage across the winding m is negligible. Therefore, the voltages on each of the windings are also negligible, and the e.m.f. measured at the two-pole terminals is $\overline{E}_{n1}$.

Second consider the effect of the equivalent noise e.m.f. $\overline{E}_{n2} = (4kT\Delta f R_{eq})^{1/2}$ of the applied amplifier element (field-effect transistor, *et cetera*), the parameter R_{eq} characterizing the noise. Again, since the feedback is assumed to be large, the voltage at the input of the amplifier is negligible, so that $\overline{E}_{n2}$ is applied to the winding n. The current through this winding is 0 due to infinite input impedance of the amplifier (and due to action of the feedback). Therefore, the currrent through the winding m is also 0; the voltage drop on R_b is 0; and the voltage across the input terminals is $\overline{E}_{n2}/n$.

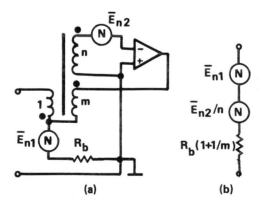

FIG. 1.39

The total noise e.m.f., therefore, is given by

$$\overline{E}_n = (\overline{E}_{n1}^2 + \overline{E}_{n2}^2 / n^2)^{1/2} = [4kT\Delta f(R_b + R_{eq}/n^2)]^{1/2}$$

compared to $(4kT\Delta fR)$ of a passive resistor having the same resistance R. Then, the equivalent temperature of the active two-pole is

$$T \frac{R_b + R_{eq}/n}{R_b(1 + 1/m)} = \frac{mT}{1+m}\left(1 + \frac{R_{eq}}{nR_b}\right)$$

For example, $R_b = 13\,\Omega$, $R_{eq} = 220\,\Omega$, $m = 0.11$, and $T = 300°\,\text{K}$ yield the equivalent noise temperature of $40°\,\text{K}$.

10. It was calculated in Sec. 1.1.6 that the system with large feedback shown in Fig. 1.40 exhibits driving-point resistance R_b/mn. Assuming, as previously, $\overline{E}_{n2} = (4kT\Delta fR_{eq})^{1/2}$, show that its equivalent noise temperature is $(1 + R_{eq}/R_b)T/n$.

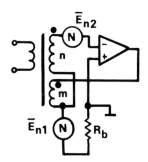

FIG. 1.40

Chapter 2
Stability and Frequency Response Constraints of Linear Systems

2.1 NYQUIST STABILITY CRITERION

2.1.1 Nyquist Diagram

Consider a rational function which is analytical on a simple closed contour in the plane of the complex variable s and inside the contour. While s makes a round trip about the contour, each multiplier of the kind $(s - s_i)$ of the function changes its argument by 2π if and only if s_i lies within the contour. In other words, each zero located inside the contour causes the locus of the function to make one revolution. Similarly, each multiplier of the form $(s - s_j)^{-1}$, with the pole s_j inside the contour, causes the locus to revolve in the opposite direction. Generally, in accordance with the argument principle [145], the number of revolutions about the origin of the locus of the function equals the difference between the numbers of the function zeros and poles within the contour.

This principle may be applied to the return difference $F = \Delta/\Delta°$ for a circuit element (recall Sec. 1.13). Using the contour composed of the $j\omega$-axis and the π-radian arc of infinite radius encompassing the right half-plane of s, the rule follows that the number of revolutions of the locus of F gives the difference between the numbers of zeros of Δ and of $\Delta°$ in the right half-plane of s.

For those circuit elements whose F has the feature of

$$\lim_{s \to \infty} F(s) = 1 \tag{2.1}$$

F does not change as s moves along the arc. Then, the rest of the contour, i.e., the $j\omega$-axis, can be considered as the contour. Because the change of the sign of ω makes F complex conjugate, the full locus of F consists of two image-symmetrical halves relating respectively to positive and negative frequencies. Hence, the locus drawn for positive ω, called the *Nyquist diagram*, makes half the number of revolutions of the whole locus.

Further, if $\Delta°$ does not possess right-hand zeros, i.e., if the system without the circuit element is stable, the Nyquist criterion follows: Δ has no right-hand zeros, i.e., the system is stable, if and only if the Nyquist diagram does not encircle the origin.

The Nyquist diagram may be regarded as a mapping of the positive $j\omega$-semiaxis of the s-plane onto the F-plane. Presuming $F(s)$ is analytical on the $j\omega$-axis, the mapping is conformal (if not, the contour may comprise small arcs to circumvent the singularities). The right half-plane vicinity of the positive $j\omega$-semiaxis plots onto the right-hand vicinity of the locus $F(j\omega)$, as shown in Fig. 2.1, and the entire first quadrant of the s-plane maps onto the interior of the Nyquist diagram.

It follows that if the Nyquist diagram encompasses the origin, some roots of the equation $F(s)=0$ must have positive real parts, i.e., the system is unstable.

The locus of $F(j\omega)$ may self-intersect as shown in Fig. 2.1. The kind of self-intersection shown in Fig. 2.2 is impossible because it contradicts the foregoing statement that the right half-plane of s falls onto the interior of the mapping.

The zeros of $F(s) = \Delta(s)/\Delta°(s)$ all map onto the origin of the F-plane. Therefore, the typical mapping of $F(s)$ represents a multifold Riemann surface, specifically in the vicinity of the origin.

More often than for F, the Nyquist diagram is used for the return ratio $T = F - 1$. In this case,

$$\lim_{s \to \infty} T(s) = 0 \tag{2.2}$$

must be held, and the point -1 serves as the critical point.

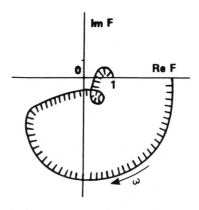

FIG. 2.1

Stability & Frequency Response Constraints

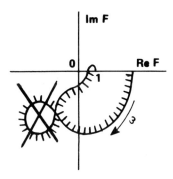

FIG. 2.2

In a system stability analysis, there are several options for selecting the circuit element whose F or T will be plotted. For example, consider the system of Fig. 2.3. It is apparently stable if either of Z_1, Z_2, Y_3, or the transistor amplification coefficient equals 0. At higher frequencies, return differences for all these circuit elements uniformly tend toward 1. Then, any one of them can be employed.

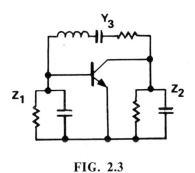

FIG. 2.3

In selecting the element, the following two considerations are of importance:

A certain minimal stability margin on the Nyquist plane for the allowable position of the Nyquist diagram must guarantee the circuit stability with given tolerances of the circuit active elements.

The effect of the system nonlinear elements on the stability conditions should be considered on the basis of the same stability margin.

These requirements are satisfied together for the single-loop feedback system comprising a sole nondynamic, nonlinear element if we use the loop return ratio as T.

2.1.2 Nyquist Diagram and Stability Margins for the Amplifier Return Ratio

The return ratio for the single-loop system of Fig. 1.4, i.e., the return ratio for w_{43}, can be used for the Nyquist criterion because the system is evidently stable without the amplifier, and, with the increase of frequency, the loop gain of any physical loop reduces uniformly to 0, i.e., (2.2) is satisfied. The transfer coefficients of the active links in the multistage forward path enter T as multipliers, thus allowing straightforward determination of the minimum stability margins for the locus of $T(j\omega)$.

The output overshoot in response to a step input must be limited in tracking systems. For this purpose, the main determinant's zeros must be located in the s-plane to the left of the rays making an angle α with the $j\omega$-axis, as shown in Fig. 2.4(b). This condition is reflected, in the following, by a restriction on the Nyquist diagram for $T(j\omega)$.

By way of the conventional procedure for an approximate analytical extension with a net of small triangles, the vicinity of the $j\omega$-axis can be found as those points of the s-plane whose images plot onto the origin of the F-plane. In practice, however, such a procedure is difficult to execute directly. On the part of the diagram facing the critical point (let us say, for $0.3 < |T| < 10$), the function $T(s)$ is typically close to a power function as^{-n} having the exponent $n \cong 1.5$. Such a function, according to Bode's phase-gain-frequency relationship, furnishes a nearly exponential frequency scale on the Nyquist diagram. Therefore, the sides of the triangles rapidly decrease along the locus. Then, since even the largest of the triangles must be small enough to avoid distortion with the transform, the smallest triangle becomes so small that the total number of the triangles required to perform the plotting becomes very large, thus obstructing the analysis.

With logarithmic variables of $\ln T$ and $\ln s$, analysis is easier. Each quadrant of the s-plane maps upon the $\ln s$-plane into a strip of the width $\pi/2$ and infinite length. The image of the first quadrant ($\omega > 0, \operatorname{Re} s > 0$) is shown in Fig. 2.5(b).

For $T(s) = as^{-n}$, $\ln T = -n \ln s + \ln a$. Therefore, linear scale of the axis $\ln \omega$ plots into a fairly linear scale of the locus $\ln T$, as seen in Fig. 2.5, so that the mapping barely distorts the shape of rather large patterns.

In the $\ln s$-plane, the straight line r shown in Fig. 2.5(b) represents the image of the ray R in s-plane. Then, as can be seen in Fig. 2.5, the phase stability margin $y\pi = -n\alpha$. Thus, increasing y reduces the overshoot.

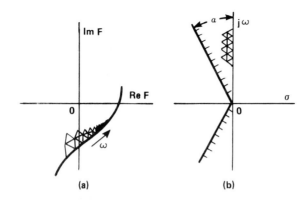

FIG. 2.4

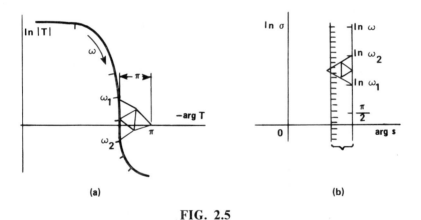

FIG. 2.5

For the responses of $T(j\omega)$ that are of practical interest, $|F|$ achieves its minimum of $\sin y\pi$ when $|T| = \cos y\pi \cong 1$, as illustrated in Fig. 2.6. With $y < 1/2$, the feedback at this frequency is positive.

To guarantee an acceptably small overshoot of the closed-loop system transmission, the value of positive feedback must be restricted, i.e., $|F|$ has to be limited from below. The inequality $|F| > \sin y\pi$ describes a disk with the radius $\sin y\pi$ centered at the critical point -1, as shown in Fig. 2.7, which the locus of $T(j\omega)$ is not allowed to penetrate.

To consider the phase and gain tolerances of the loop transmission function, the region around the critical point is often shaped as shown in Fig. 2.7(a) by the segment *klmn* on the *T*-plane, or, identically, by the rectangle *klmn* on the Nichols plane of Fig. 2.7(b).

The amplitude stability margins x and x_1 can be chosen either symmetrically ($x = x_1$) or not, depending on the types of nonlinear links in the loop. For single-loop systems with memoryless nonlinear links, the conventional

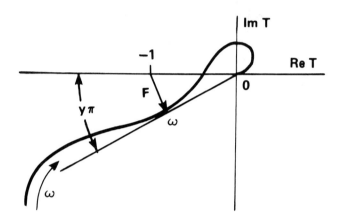

FIG. 2.6

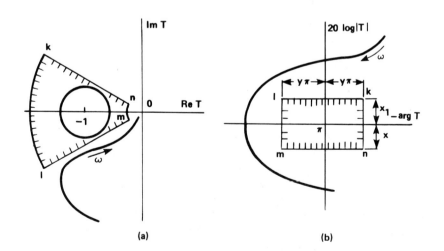

FIG. 2.7

Bode stability margins shown in Fig. 2.8 (with $x_1 = \infty$) guarantee asymptotic global stability, to be discussed further in Chapter 3. Common values are $y = 1/6$ to $1/4$, and $x = 6$ to $10\,\text{dB}$.

The plant frequently comprises, in addition to the actuator and a linear link Q, a path β_l of local feedback as depicted in Fig. 2.9. Such a plant is sometimes inherently unstable, the hodograph of $\beta_l Q$ encircling the critical point -1. Making such a plant stable and controlled is achieved by the loop of common feedback. The stability analysis of the system is simplified when employing the return ratio $T = Q(\beta_l + M)$, measured at the input of the actuator, to consider the variations of this link transmission caused by overloading.

The stability margin may not be narrower than that dictated by the requirement to bound the forced oscillation, which is a topic to be discussed in Chapter 4.

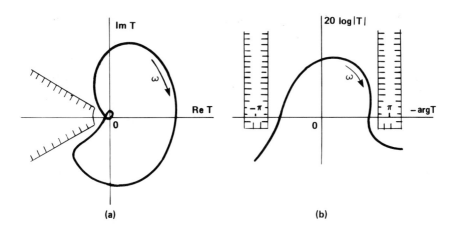

FIG. 2.8

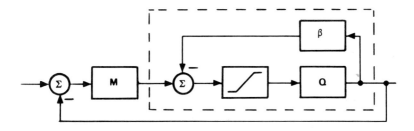

FIG. 2.9

2.1.3 Two-Poles' Coupling

Zeros of a driving-point immittance (i.e., impedance or admittance) relate to the eigenvalues of the circuit. Therefore, when coupled to a zero immittance two-pole, the two-pole is unstable if and only if its immittance possesses zeros in the right half-plane of s. Similarly, when loaded at an infinite immittance two-pole, a circuit oscillates if and only if its driving-point immittance possesses right-hand poles.

If the immittance of the two-pole has no right-hand zeros, it is apparently stable when connected to a two-pole having an immittance module which is sufficiently small. If the immittance of the two-pole possesses both zeros and poles in the right half-plane, it may be that there is no passive load for making the circuit stable.

As mentioned in Sec. 1.1.3, the equation describing two-pole coupling can be put into the form of the block diagram of a feedback system in Fig. 1.1 with return ratio $T = W_1/W_2$. The stability of the circuit depends on whether the return difference $F = T + 1 = W_1/W_2 + 1$ has right-hand zeros. The Nyquist criterion could be used if $\lim_{\omega \to \infty} T = 0$; otherwise, it is always possible to redesignate W_1 as W_2 and vice versa.

It follows from the principle of argument that the number of right-hand poles of $W_1 + W_2$, i.e., of F, equals the sum of the number of the right-hand poles of W_1 and the right-hand zeros of W_2, after subtracting twice the number of revolutions of the Nyquist diagram for W_1/W_2 about the point -1.

The Nyquist criterion is often used in simplified form, giving a sufficient stability criterion. When, for example, stability of an active two-pole Z_1 loaded into an unknown passive two-pole Z_2 is required, it is sufficient that at all frequencies $\operatorname{Re} Z_1 > 0$ so that $|\arg Z_1/Z_2| < \pi$, preventing the Nyquist diagram for Z_1/Z_2 from encircling the critical point -1.

2.1.4 Reflection Coefficients

Consider the product

$$\Gamma_1 \Gamma_2 \tag{2.3}$$

where

$$\Gamma_1 = \frac{Z_1 - Z_c}{Z_1 + Z_c}$$

and

$$\Gamma_2 = \frac{Z_2 - Z_c}{Z_2 + Z_c}$$

stand for the reflection coefficients calculated for the two-poles having impedances Z_1 and Z_2, with respect to the given passive impedance Z_c. It is easy to see that the poles of expression (2.3) are produced by the roots of equation $(Z_1 + Z_c)(Z_2 + Z_c) = 0$, and the zeros by the roots of equation $Z_1 + Z_2 = 0$.

Without contradicting reality, we may assume that each of the two-poles includes a short transmission line connecting its interior with the terminals, the lines' attenuation increasing indefinitely with frequency, as shown in Fig. 2.10(a). Then, at very high frequencies, Z_1, Z_2, and Z_c become equal to the characteristic impedance of the line, thus causing Γ_1 and Γ_2 to approach 0.

Suppose further that each of the two-poles is stable, being loaded at a two-pole with the impedance Z_c, i.e., $Z_1 + Z_c$ and $Z_2 + Z_c$ have no zeros in the right half-plane of s. In this case, the condition of applicability of the Nyquist criteria for $\Gamma_1 \Gamma_2$ is satisfied. Then, the sum $Z_1 + Z_2$ has no right-hand zeros, i.e., interconnection of the two-poles with these impedances is stable if and only if the Nyquist diagram for $\Gamma_1 \Gamma_2$ does not encircle the point -1, i.e., if and only if the equivalent feedback system of Fig. 2.10(b) is stable.

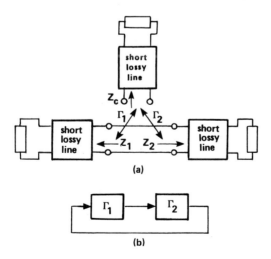

FIG. 2.10

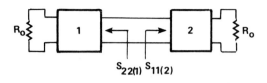

FIG. 2.11

If the two-ports in Fig. 2.11 have standard termination (let us say, 50 Ω) from outer sides, the reflection coefficients Γ_1 and Γ_2 represent S-parameters of the two-ports, correspondingly $s_{22(1)}$ for the first of the two-ports and $s_{11(2)}$ for the second, and encirclement of the origin by the Nyquist diagram drawn for the function $s_{11(2)}s_{22(1)}$ gives the system stability criterion, under the condition that this product vanishes with frequency rising indefinitely.

In microwave engineering, a two-port is called conditionally stable at a specified frequency if the real parts of its input and output impedances are positive for some specific positive-real source and load impedances, and unconditionally stable if they are positive for any combination of positive-real source and load. When the gain of a microwave amplifier is maximized by making its input impedance complex conjugate to the source impedance, and the output impedance to the load impedance, and the amplifier is not unilateral, the conditional stability condition is expressed as $(1+|s_{11}s_{22}-s_{12}s_{21}|-|s_{11}|^2-|s_{22}|^2)/2|s_{12}s_{21}|>1$; the left side of this inequality is known as the stability factor [133]. The stability analysis employing the stability factor and the Smith chart serves the design problems of single-stage amplifiers and oscillators well, but is inexpedient for the design of multistage, globally stable systems with maximized feedback.

Note that in the rest of the book we accept different definitions of conditional and unconditional stability, which are standard in the mathematical theory of stability.

2.1.5 Examples and Exercises

1. Figure 2.12 depicts a feedback system composed of three identical amplification stages. The system is unstable if the Nyquist diagram for the amplifier return ratio encloses the critical point -1, i.e., if the phase lag of each stage exceeds 60° at the crossover frequency f_b at which $|T|$, decreasing with frequency, reaches 1. The system faces instability if at zero frequency $|T|=(\cos 60°)^{-3}=8$.

The system remains stable with much larger loop gain at low frequencies if one of the shunting capacitances is many times increased, cutting off the loop gain with the slope $-6\,\text{dB/oct}$. This leads to the equality $|T| = 1$ at a frequency smaller than that at which two other stages together contribute $\pi/2$ to the loop phase lag. This simple method of stable feedback system design, however, narrows the frequency band of large feedback. A more economical design is considered in Sec. 2.4.

2. An amplifier with a three-link RC-filter in the feedback path shown in Fig. 2.13 may represent an oscillator, i.e., its return ratio T could encircle the critical point. If, however, either one of the resistors $R_e = \infty$, then $|\arg T| < \pi$ and the circuit is stable. It is also evidently stable if $R_e = 0$.

Alternatively, the system may be viewed as the connection of resistor R_e to the active two-pole Z, which represents the rest of the circuit. Since Z is stable while loaded at 0 and ∞, it has neither poles nor zeros in the right half-plane of s. Therefore, we can use the Nyquist diagram for the return ratio related to the impedance Z or admittance $1/Z$. The system is, therefore, stable if and only if the hodograph for the return ratio as Z/R or R/Z does

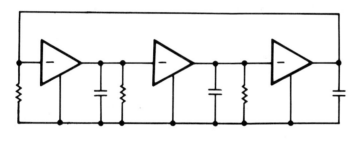

FIG. 2.12

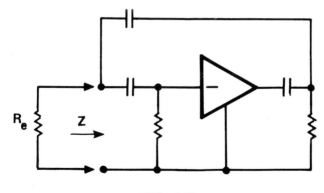

FIG. 2.13

not encircle the point -1. Gradually changing the gain of the amplifier renders the situation of facing the oscillation, when $Z = -R_e$. In this case, the return ratio for the amplifier also ought to be -1 (recall Eq. (1.29)).

It is instructive to determine which method of stability analysis better suits the system that has been designed to be stable with an amplifier whose gain accepts any value from a certain nominal value down to 0, or when the resistance of a two-pole varies over a certain interval. Such a problem arises, in particular, when either the amplifier or the two-pole represents a nonlinear device and if for a rough analysis the system is considered as quasilinear, replacing the nonlinear element by an equivalent linear element that varies with the signal level.

If only the amplifier gain is changing, then using the Nyquist diagram for the amplifier return ratio simplifies the analysis — because as the gain decreases, the diagram merely shrinks down to the origin, its shape preserved. Hence, it is possible to see from the Nyquist diagram for the nominal gain value whether the diagram will encompass the critical point while shrinking. Conversely, if R_e is variable, the single diagram either for $T = Z/R_e$ or for $T = R_e/Z$ satisfies the analysis.

3. Consider a realization of a NIC by the system of Fig. 1.23(b). At frequencies from 0 to ω_1, the circuit closely approximates the NIC. Beyond this frequency, the feedback in the system dies down and the parameters of the two-port become frequency-dependent. It can be seen that if the left terminals are shorted or the right terminals are open, the feedback signal is initially inverted in the loop, and if the feedback is properly designed, the system is stable. Therefore, the impedance measured at the left terminals and the admittance measured from the right terminals, have no right-hand zeros. Hence, the system is stable being open at the left terminals or shorted at the right. Thus, while loading the NIC at any passive impedance from the opposite side, the locus of input impedance (or the output admittance from the right side) does not encompass the origin.

Figure 2.14 represents the hodograph of the input impedance of a NIC measured while loading the opposite port at 1 kΩ. From which side was this impedance measured? How can the circuit stability be proved using the Nyquist diagrams for driving-point impedances? Does the diagram provide a basis for the judgement about what would happen if the gain of the amplifier or the load resistance were reduced?

4. Next consider the stability problem in a system with three variable parameters. Figure 2.15(a) shows a resonance stage of amplification with the feedback caused by a small stray capacitance. Its impedance jX may be considered as constant over the rather narrow frequency range of possible instability, which may only occur close to the resonance frequency of the

Stability & Frequency Response Constraints

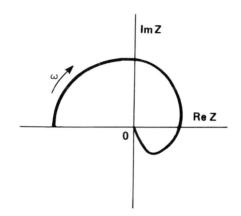

FIG. 2.14

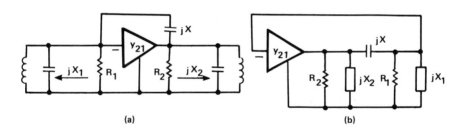

FIG. 2.15

input and output resonant tanks. The tank impedances Z_1 and Z_2 are represented by a parallel connection of branches with real (R_1 and R_2) and imaginary (X_1 and X_2) impedances. R_1 and R_2 are considered stable in time, but X_1 and X_2 could accept any value due to mistuning of the resonance contours. Figure 2.15(b) represents the same circuit redrawn so as to clarify what constitutes the feedback path.

Normally, $|X| \gg |Z_1|, |X| \gg |Z_2|$; then

$$T \cong y_{21} Z_2 Z_1 / jX$$

where y_{21} stands for the output/input real transconductance of the amplifier element.

To guarantee stability of the system with any transconductance smaller than the nominal value of y_{21} and with any X_1, X_2, we require that $|T|$ be less than 1 in all cases when

$$\arg T = \pi$$

i.e., when $\arg Z_1 Z_2 = \pi/2$. It is easy to prove that with variation of $\arg Z_1$, under the above condition, $|T|$ is maximal when $\arg Z_1 = \arg Z_2 = \pi/4$. Concentrating on this worst case,

$$T = -\frac{y_{21}}{X} \frac{R_1 R_2}{2} \qquad (2.6)$$

For the system to be stable with Bode amplitude stability margin x dB, the modulus of (2.6) should be smaller than

$$\sigma = 10^{-x/20}$$

and, therefore, finally,

$$R_1 R_2 \leq \frac{2\sigma X}{y_{21}} \qquad (2.7)$$

For a single-stage amplifier, it is sufficient to use $x = 8$ dB so that $\sigma = 0.4$.

This is the same result as found in [163] using driving-point impedance analysis. The advantage of analysis by the return ratio for the amplifier is that the methods described in the next chapter can be applied to prove the system global stability easily.

5. A symmetrical active two-port with the transmission constant g_c and a positive-real image impedance z_c, loaded from the left side at the impedance Z_a and from the right side at Z_b, is shown in Fig. 2.16. Since it is known [60] that

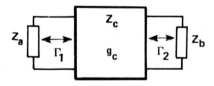

FIG. 2.16

$$\Gamma_{in} = \frac{Z_{in} - z_c}{Z_{in} + z_c} = \frac{Z_b - z_c}{Z_b + z_c} \exp(-2g_c) = \Gamma_b \exp(-2g_c),$$

application of the (2.3) to the left-hand terminals of the two-port gives

$$T = -\Gamma_a \Gamma_b \exp(-2g_c)$$

where $\Gamma_a = (Z_a - z_c)/(Z_a + z_c)$. The circuit is stable if the Nyquist diagram for this T does not encircle the point -1. The equivalent feedback system is shown in Fig. 2.17.

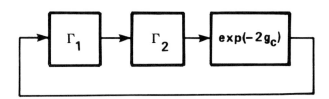

FIG. 2.17

6. The bilinear transform $\Gamma_b = (Z_b - z_c)/(Z_b + z_c)$ maps the right half-plane of Z_b, i.e., all passive load impedances, into the disk shown in Fig. 2.18(b). Here, the imaginary axis of Z_b-plane is transformed into a circumference symmetrically situated with respect to the imaginary axis and passing through the points 1, -1, and $j\cot(\psi/2 + \pi/4)$, where $\psi = \arg z_c$. It can be seen that with any $\arg \Gamma_b$, the maximum of $|\Gamma_b|$ is achieved at the boundary

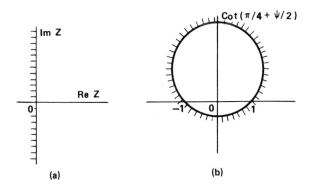

FIG. 2.18

of the disk (this follows also from the theorem of the maximum modulus of an analytical function [145]). Therefore, if the circuit of Fig. 2.16 is supposed to be stable with any positive-real Z_a or Z_b, it is sufficient to perform the analysis for purely reactive loads.

7. Consider a two-port which is to be designed to be stable while being embedded between the load Z_L and the source Z_s, with the areas of possible Z_L and Z_s given. The stability analysis of such a two-port is often approached using S-parameters of the two-port, especially in active microwave circuit design.

On the plane of the input impedance Z_1 and reflection coefficient with respect to nominal impedance R_0 (i.e., on the Smith chart), the area of possible loads at a certain frequency is reflected into area L. The area of acceptable source impedances Z_s is reflected into area S. If these areas do not intersect, at any frequency, then the ratio Z_s/Z_1 never equals -1, and the Nyquist diagrams for any of pair Z_s and Z_R have the same number of revolutions around the critical point. Then, the number of right-hand zeros of the system's main determinant is the same for all pairs of Z_s and Z_L (excluding zero probability cases of simultaneously changing numbers of poles and zeros).

Then, in order to prove the circuit unconditional stability (in the sense accepted in microwave engineering), it only remains to be shown that at any specific pair Z_s and Z_L the system is stable, which is fairly simple.

If the areas of allowable Z_s and Z_L are bounded by circumferences, the areas S and L are also shaped as certain disks, due to the property of bilinear functions [23,145], as illustrated in Fig. 2.19. In particular, if Z_L is arbitrarily passive, the radius of L is expressed through the S-parameters of the two-port as $|s_{11}s_{22}/(|s_{22}|^2 - |\Delta|^2)|$, its center located at $(s_{22} - \Delta s_{11}^*)^*/(|s_{22}|^2 - |\Delta|^2)$, where $\Delta = s_{11}s_{22} - s_{12}s_{21}$, and the asterisk designates complex conjugate values.

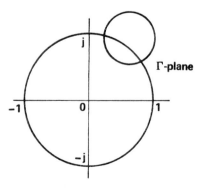

FIG. 2.19

2.2 INTEGRAL CONSTRAINTS

2.2.1 Minimum Phase Functions

If the function $\theta(s)$ possesses no poles on the $j\omega$-axis (the logarithmic singularities are permitted), then for any frequency ω_0 including ∞,

$$\lim_{\omega \to \omega_0} (\omega - \omega_0) \, \theta(j\omega) = 0 \tag{2.8}$$

If, further, $\theta(s)$ is analytical in the right half-plane of s, the contour integral of θ around the right half-plane equals 0. Bode named such functions *minimum phase* (MP) because any other physically realizable function having the same real-part frequency response possesses an additional negative-imaginary component, called *nonminimal phase shift* (NPS).

Many network functions belong to the class of MP functions, such as logarithms of the transmission function of a passive ladder network. The driving-point impedance of a passive network is MP if this impedance is limited over the $j\omega$-axis [23,83].

As will be shown further, the phase delay in the feedback loop limits the available feedback. Therefore, it is desirable that the logarithmic transfer function of the feedback loop links be MP, or at least that their NPS be small.

2.2.2 Integral of the Real Part

The Laurant expansion at $s \to \infty$ for a MP function $\theta(s)$:

$$\theta(s) = A_\infty - \frac{B_\infty}{s} - \frac{A_1}{s^2} - \frac{B_1}{s^3} - \frac{A_2}{s^4} \cdots$$

converges over the right half-plane of s. On the $j\omega$-axis, it accepts the form:

$$\theta(j\omega) = A_\infty + j\frac{B_\infty}{\omega} + \frac{A_1}{\omega^2} + \cdots \tag{2.9}$$

The function $\theta - A_\infty$ is MP as well. Then, the contour integral of $\theta - A_\infty$ around the right half-plane of s equals 0. The contour of integration may be viewed as composed of the $j\omega$-axis completed by an π-radian arc of infinite radius. The integral along the arc equals πB_∞ [23,145]; the integral along the entire jω-axis equals twice the integral of the even part of the integrand, i.e., of $A - A_\infty$, along the positive semiaxis. From here,

$$\int_0^\infty (A - A_\infty)\, d\omega = -\frac{\pi B_\infty}{2} \tag{2.10}$$

By specifying the meaning of A and B entering (2.10), several important feasibility conditions are generated.

First, assign to θ the meaning of $\ln F$. At higher frequencies,

$$\theta = \ln(1 + T) \cong \ln\left(1 + \frac{a}{(j\omega)^n}\right) \cong \frac{a}{(j\omega)^n}$$

where a is a constant. We see that $A_\infty = 0$. For the most common case of $n \geq 2$, the coefficient $B_\infty = 0$, and

$$\int_0^\infty \ln|F|\, d\omega = 0 \tag{2.11}$$

The integrand in (2.11) is positive when the feedback is negative, and *vice versa*. Therefore, the integral over the frequency region where the feedback is negative equals, with opposite sign, the integral over the range of positive feedback. It follows that if a system possesses negative feedback in a certain frequency range, there should exist another frequency range where the feedback is positive. The larger the negative feedback and its frequency range, the larger is the area of positive feedback.*

Second, let $\theta(j\omega) = Z(j\omega) = R(\omega) + jX(\omega)$ stand for the impedance of a parallel connection of a capacitance C and a two-pole having impedance Z', as shown in Fig. 2.20. Z' is assumed as limited at all frequencies and not reducing to 0 at infinite frequency. Then, at higher frequencies, Z is approximately expressed as

$$Z = \frac{1}{j\omega C}$$

Comparing this formula with (2.9) gives $A_\infty = 0$, $B_\infty = 1/C$, and from (2.10) we have the equation called the integral of resistance:

$$\int_0^\infty R\, d\omega = \frac{\pi}{2C} \tag{2.12}$$

---

*This theorem should be remembered while discussing the criteria of optimal feedback [81].

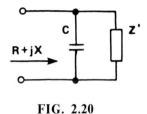

FIG. 2.20

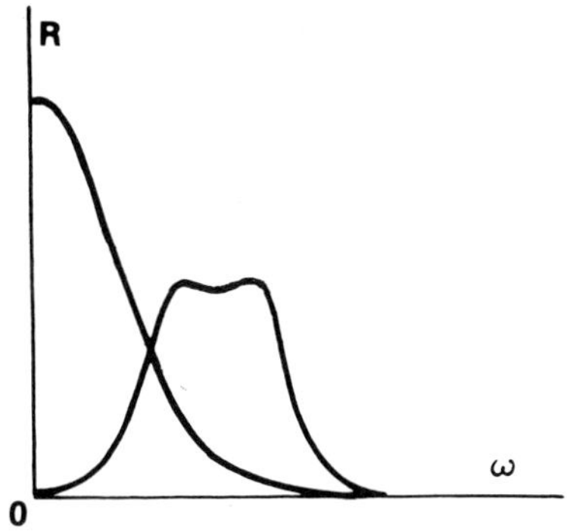

FIG. 2.21

Thus, the area under the frequency response of the resistance is only determined by the parallel capacitance, and not by the impedance Z'. This is illustrated by the frequency responses of R in Fig. 2.21 for the two-poles of Fig. 2.22.

Relation (2.12) as well as the integral of the real part of the admittance of the dual circuit of Fig. 2.23, which is

$$\int_0^\infty \operatorname{Re} Y(\omega) \, d\omega = \frac{\pi}{2L} \qquad (2.13)$$

are widely applied to evaluation of the available bandwidth-performance product in systems where the stray reactive element, C or L, becomes critical, in particular, in the input and output circuits of wideband amplifiers.

In Fig. 2.24(a), the output of an amplifier represented by the current source I shunted by the stray capacitance C is connected to the load R_L through an LC-corrector. Due to the conservative (nondissipative) character of the circuit, the power arriving at the load U^2/R_L equals $I^2 R$, where U is the

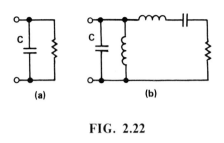

FIG. 2.22

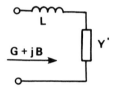

FIG. 2.23

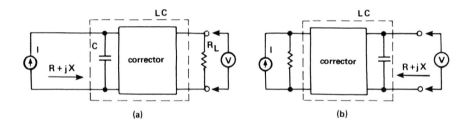

FIG. 2.24

voltage across the load and R is the real part of the impedance facing the current source. In order to increase the output voltage $U = I\sqrt{RR_L}$, the resistance R must be maximized within the limit given by (2.12).

Consider next the input circuit depicted in Fig. 2.24(b), where I and R_L represent the source of signal, and the stray capacitance C shunts the amplifier input. The circuit differs from that of Fig. 2.24(a) in only the direction of signal propagation. Application of the reciprocity theorem [23] proves that the formula $U = I\sqrt{RR_L}$ is valid for the circuit of Fig. 2.24(b) as well. Thus, the value of R constrained by (2.12) limits the gain of the input circuit and, therefore, limits from below the minimum available noise figure.

To maximize the gain of the input and output circuits over the band of operation Δf, the value of R must be at the limit allowable according to (2.12),

$$R_{max} = \frac{1}{4C\Delta f} = \frac{\pi}{2} X \qquad (2.14)$$

where $X = 1/2\pi C\Delta f$. This is illustrated in Fig. 2.25. It is noteworthy that only R, but neither $|Z|$ nor the reflection coefficient, remain constant over the band. For this reason, such a frequency response of R might not suit the output of a power amplifier where the considerations of minimizing the output voltage swing and nonlinear distortions are of importance.

In Fig. 2.25, outside the working band, $R = 0$. Hence, the LC-corrector together with C constitute an ideal filter of infinite selectivity. The number n of the elements is infinite in the ideal filter. When n is finite, a part of the area under the frequency response of R falls outside the working band, making the available R smaller within the band.

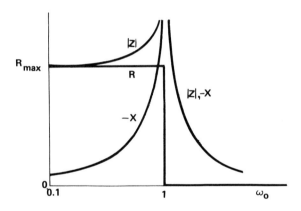

FIG. 2.25

Third, consider the reflection attenuation A_r for the two-pole of Fig. 2.20. For the sake of simplicity, normalize the nominal value of R to be 1. If the reflection coefficient $\Gamma = (1 + Z)/(1 - Z)$ possesses no zeros in the right half-plane of s, i.e., if $1 - Z$ has no right-hand zeros (the LC two-port can be designed this way [23]), then, the function $\ln \Gamma$ can be employed as θ in (2.10). In this case, at higher frequencies,

$$\theta = \ln \frac{1 - j/\omega C}{1 + j/\omega C} = j \frac{2}{\omega C}$$

so $A_\infty = 0$ and $B_\infty = -2/C$. Therefore, (2.10) yields the integral constraint for the reflection attenuation $A_r = \text{Re}\,\theta$ in the form of

$$\int_0^\infty A_r \, d\omega = \pi/C \tag{2.15}$$

For the dual circuit, similarly,

$$\int_0^\infty A_r \, d\omega = \pi/L \tag{2.15}$$

If the task for the circuit synthesis is stated to achieve maximum available A_r over the working band, from (2.15), the maximum available A_r is

$$A_{r\,max} = \pi X_c / R_0 \tag{2.16}$$

and, for the dual circuit,

$$A_{r\,max} = \pi R_0 / X_L \tag{2.17}$$

where, as before, $X_c = 1/2\pi \Delta f C$ and $X_L = 2\pi \Delta f L$.

These relations plotted in Fig. 2.26 give the maximum admissible values for C (or L) that can be tolerated while preserving the prescribed value of $A_{r\,max}$ over the band of Δf, with the number n of the elements in the corrector being infinite. The dependencies for finite n show the maximum of the minimum A_r attainable within the band [0,1] with a ladder-type (polynomial) filter having an equiripple response of R [44].

In a lossless two-port, the incident power divides between that which is transmitted to the load and that which is reflected back, so the following equality occurs:

$$1 = 10^{-A} + 10^{-A_r}$$

If the response for A is equiripple (Chebyshev), and the reflection attenuation A_r is large (more than 15 dB), the frequency response for A_r is nearly equiripple as well. Therefore, although special polynomial filters approximating A_r in the Chebyshev sense were developed [44], using standard Chebyshev filters [61,83] is quite appropriate. Normalized element values for the Chebyshev filters of Fig. 2.27 are presented in Table 2.1. In order to obtain the element values, normalized inductances have to be multiplied by $R_0/2\pi f_0$, and normalized capacitances by $1/2\pi f_0 R_0$, where R_0 is the termination resistance, and f_0 is the cut-off frequency.

These filters employ lumped elements. Filters built with distributed parameter elements, like sections of long lines, yield poorer band utilization because of the distributed capacitance (or inductance, for the dual problem)

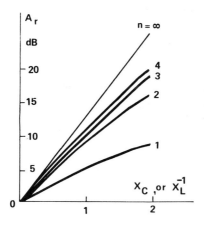

FIG. 2.26

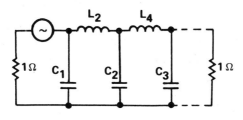

FIG. 2.27

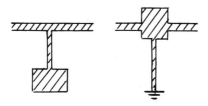

FIG. 2.28

of the first section of the filter. This is why the design of microwave amplifiers benefits from using resonators like those shown in Fig. 2.28 with a narrow printed circuit or a wire bond as an inductor, instead of quarter-wave resonant stubs.

2.2.3 Integral of the Imaginary Part

If θ is MP, then θ/s is analytical in the right half-plane of s and on the $j\omega$-axis, except for the origin. So, if the origin is avoided along the π-radian arc of infinitesimal radius, then the integral of θ/s about the contour encompassing the right half-plane equals 0. The integral along the small arc equals πA_0; along the large arc it is πA_∞. Therefore, the integral of the even part of θ/s along the $j\omega$-axis equals

$$\int_{-\infty}^{\infty} \frac{B}{\omega} d\omega = \pi(A_\infty - A_0)$$

i.e.,

$$\int_{-\infty}^{\infty} B(u)\, du = \pi(A_\infty - A_0)/2 \tag{2.18}$$

where $u = \ln \omega$. This relation is known as the phase integral.

Corollary: For two frequency responses A' and A'' joining together at higher frequencies, the difference in the related phase integrals is $(A_0' - A_0'')\pi/2$.

With application to the loop gain of a feedback system, this corollary signifies that given the loop transmission function at higher frequencies, an increase in the loop gain in the band of operation implies increasing the area below the frequency response of the loop phase lag. However, the phase lag is restricted by stability condition requirements. Hence, the feedback in the band of operation is limited.

Another important relation between the real and imaginary components results from the integral of $(\theta - A_\infty)\sqrt{sW}$ equaling 0 around the same contour; here, W is a reactance function [23,83], i.e., an impedance function of a reactance two-pole. On the $j\omega$-axis, W is purely imaginary, either positive or negative. The function $\sqrt{sW}$ is, therefore, either purely real or purely imaginary, alternately at the adjoint sections over the $j\omega$-axis, and possesses branch points at the joints of these sections. The sign of the radix at the sections must be chosen so that the whole contour of integration belongs to only one of the Riemann folds. On this contour, the function Re $W(j\omega)$ must be even and Im $W(j\omega)$ odd.

For those values of W that reduce to s at higher frequencies, the integrand decreases with s at least as s^{-2}, and the integral along the large arc vanishes. Then, since the total contour integral is 0, the integral along the $j\omega$-axis equals 0 as well. Its real part is certainly 0,

$$\int_{-\infty}^{\infty} \operatorname{Re} \frac{\theta - A_\infty}{\sqrt{sW}} \, d\omega = 2 \int_{0}^{\infty} \operatorname{Re} \frac{\theta - A_\infty}{\sqrt{sW}} \, d\omega = 0 \qquad (2.19)$$

If, in particular, $W = (1 - s^2)/s$ and $\sqrt{sW} = \sqrt{1 - \omega^2}$ is real for $|\omega| < 1$ and imaginary otherwise, then (2.19) yields

$$\int_{0}^{1} (A - A_\infty)[1 - \omega^2]^{-1/2} \, d\omega + \int_{1}^{\infty} B[1 - \omega^2]^{-1/2} \, d\omega = 0$$

i.e.,

$$\int_{\omega=0}^{\omega=1} (A - A_\infty) \, d \arcsin \omega = -\int_{1}^{\infty} B[1 - \omega^2]^{-1/2} \, d\omega \qquad (2.20)$$

With application to a single-loop feedback system, (2.20) means that reshaping the loop gain response in the working band $\omega \le 1$ does not affect the available loop phase lag at frequencies $\omega \ge 1$ as long as the area of the gain plotted against $\arcsin \omega$ is preserved. Such reshaping does not interfere with the available stability margins. This fact permits us to accept the value A_0 of the available feedback, which is flat over the working band, as a figure of merit, even for systems whose loop gain is not required to be flat.

2.2.4 Examples and Exercises

1. Stray capacitance $C = 0.8 \, \text{pF}$ shunts the input of a field-effect transistor (FET) amplifier, as shown in Fig. 2.29(a). The internal resistance of the signal source is 50 Ω. The gain should be flat within 0.3 dB from 1 GHz to 1.5 GHz. Find the maximum achievable transformation ratio for the input transformer (in order to attain the minimal noise figure).

Equation (2.14) gives 267 Ω as the maximal resistance to be faced by the FET input. Therefore, the maximum achievable transformation ratio is $(267/50)^{1/2} = 2.3$. If the LC-corrector represents a Chebyshev third-order filter having 0.3 dB attenuation ripples in the working band, Table 2.1 (Sec. 2.2.2) gives the filter terminating resistance 220 Ω for 0.5 GHz bandwidth and 0.8 pF input capacitance, so the transformation ratio is $(220/50)^{1/2} = 2.1$. The filter after bandpass transformation into the working band is shown in Fig. 2.29(a).

2. Design an unbalanced-to-balanced transformer (balun) to serve as an input circuit for a push-pull feedback amplifier. The reflection attenuation of the transformer loaded at 75 Ω, as shown in Fig. 2.29(b), should be more than 20 dB over the working band of 50 to 700 MHz.

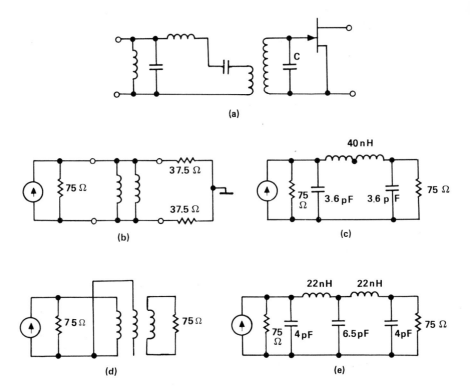

FIG. 2.29

TABLE 2.1

A (dB)	n	C_1	L_2	C_3	L_4	C_5	L_6	C_7
.18	3	1.19	1.15					
	4	1.04	1.48	1.48	1.04			
	5	1.30	1.35	2.13	1.35	1.30		
	6	1.16	1.53	1.84	1.84	1.53	1.16	
	7	1.34	1.39	2.24	1.51	2.24	1.39	1.34
.28	3	1.35	1.41	1.35				
	4	1.15	1.51	1.51	1.15			
	5	1.46	1.31	2.28	1.31	1.46		
	6	1.28	1.53	1.88	1.88	1.53	1.28	
	7	1.49	1.34	2.39	1.45	2.39	1.34	1.49

Due to a large related bandwidth, we consider low-frequency and high-frequency regions separately. Five turns in each winding on a toroidal core (bead) with magnetic permeability of 250 produce enough inductance to satisfy the requirement for 50 MHz. The wires of the primary and secondary were twisted together to reduce stray inductance that, however, still constitutes 40 nH (total value for both windings, as shown in Fig. 2.29(c)).

In order to compensate as large a part of this stray inductance as possible, the input and output of the transformer is worth shunting with capacitances of maximum permissible values. Equation (2.14) gives 5.1 pF for the ideal case of infinite number of elements filter. For a third-order polynomial Chebyshev filter having 20 dB reflection attenuation, this capacitance should be 1.6/2.25 times smaller, as can be seen in Fig. 2.26, so they should be of 3.6 pF, as indicated in Fig. 2.29(c). The values of the elements of this filter follow from Table 2.1 (Sec. 2.2.2): the capacitances are each $1.19/(2\pi 75/7 \cdot 10^8) = 3.6 \cdot 10^{-12}$ F (as we already knew, the serial inductance is $1.15 \cdot 0.75/(2 \cdot 0.7 \cdot 10^8) = 19 \cdot 10^{-9}$ H). Thus, the stray inductance of 40 nH is too high to fit the filter, and has to be substantially reduced. This was done by breaking the stray inductance into two halves — of the primary and the secondary windings, respectively — by connecting a shunting capacitance to the point where they meet, as shown in Fig. 2.29(e), thus changing the higher-frequency equivalent circuit to a polynomial filter of the fifth order. The capacitor was physically implemented by adding a third (idle) winding twisted together with the primary and secondary to form a triple wire, and grounded from the side opposite to the grounding of the primary, as shown in Fig. 2.29(d). The capacitances calculated with the help of Table 2.1 correspondingly constitute 4 pF; 6.5 pF; and 4 pF, and the inductances, each of 22 nH, now become realizable.

Although the equivalent circuit of Fig. 2.29(e) only approximates the real distributed parameter circuit, the above considerations form the approach to transformer design that will show Chebyshev performance after proper tuning. Such transformers are normally tuned during mass production by separating or compacting the turns of the windings under the control of a network analyser.

An added benefit, which serves to reduce the second-order nonlinear products in a push-pull amplifier, is superior balancing relative to ground. This is a result of the symmetry in the array of stray capacitances between the windings in the circuit of Fig. 2.29(d).

2.3 PHASE-GAIN RELATIONS

2.3.1 General Relation

Define the frequency at which the phase shift is of interest as ω_c, and $\theta(j\omega_c)$, as $\theta_c = A_c + jB_c$. Consider the function of s:

$$(\theta - A_c)\left(\frac{1}{s/j - \omega_c} - \frac{1}{s/j + \omega_c}\right) = (\theta - A_c)\frac{2\omega_c}{-s^2 - \omega_c^2} \tag{2.24}$$

It is analytical in the right half-plane of s and on its boundary except at the points $-j\omega_c$ and $j\omega_c$. Therefore, its integral taken about the contour shown in Fig. 2.30 equals 0. The contour consists of four parts, the integrals along three of which are easy to calculate, as will be shown below.

The integral along the arc of infinite radius equals 0 because of the term

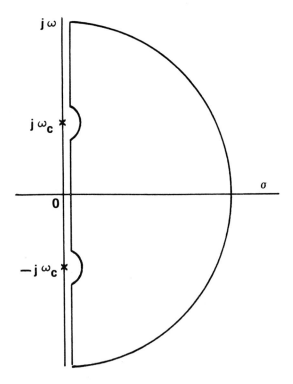

FIG. 2.30

s^2 in the denominator of the integrand.

As s approaches $j\omega_c$, then $\theta - A_c$ approaches B_c, and the second multiplier in the left side of (2.24) tends toward $1/(s/j - \omega_c)$. Then, the integral along the infinitesimal arc centered at $j\omega_c$ equals

$$\oint B_c \frac{1}{s/j - \omega_c} ds = -\pi B_c$$

The integral along the second small arc equals $-\pi B_c$ as well.

Next, setting the sum of all four components of the integral equal to 0, we see that the integral along the $j\omega$-axis equals $2\pi B_c$. Neglecting the odd component of the integrand whose integral disappears within symmetrical boundaries, we have

$$\int_{-\infty}^{\infty} (A - A_c) \frac{2\omega_c}{\omega^2 - \omega_c^2} d\omega = 2\pi B_c$$

and, finally,

$$B_c = \frac{2\omega_c}{\pi} \int_0^{\infty} (A - A_c) \frac{d\omega}{\omega^2 - \omega_c^2} d\omega \qquad (2.25)$$

Corollary 1: If A is a constant, $B_c = 0$.

Corollary 2: Consider a lowpass system with the frequency response of attenuation A shown in Fig. 2.31. At ω_c much lower than the cut-off frequency ω_0, the phase shift

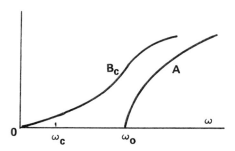

FIG. 2.31

$$B_c \cong \frac{2\omega_c}{\pi} \int_{\omega_0}^{\infty} \frac{A - A_c}{\omega^2} d\omega \qquad (2.26)$$

is proportional to the frequency ω_c.

Equation (2.25) may be modified by integration by parts, since

$$\int \frac{A - A_c}{\pi} \frac{2\omega_c}{\omega^2 - \omega_c^2} d\omega = \int U\, dV = UV - \int V\, dU \qquad (2.27)$$

where we denote:

$$U = \frac{A - A_c}{\pi}$$

$$V = \int \frac{2\omega_c d\omega}{\omega^2 - \omega_c^2} = -\ln \coth \frac{u}{2}$$

$$u = \ln \omega / \omega_c$$

Thus, expression (2.25) equals

$$B_c = \frac{-1}{\pi} \left[(A - A_c) \ln \coth \frac{u}{2} \right]_{-\infty}^{\infty} + \frac{1}{\pi} \int_{-\infty}^{\infty} \frac{dA}{du} \ln \coth \frac{u}{2} du \qquad (2.28)$$

The left side of the equation is real. Hence, the imaginary components on the right side must disappear after summing. We can, therefore, count the real components only. Because

$$\ln \coth \frac{-u}{2} = j\pi + \ln \coth \frac{u}{2}$$

replacing $\ln \coth(u/2)$ by $\ln \coth|u/2|$ does not change the real components of (2.28), and is therefore permitted. After replacement, the function in square brackets becomes even and disappears because of symmetrical limits. Then, finally,

$$B_c = \frac{1}{\pi} \int_{-\infty}^{\infty} \frac{dA}{du} \ln \coth \frac{|u|}{2} du \qquad (2.29)$$

As can be seen from this equation, the phase shift B_c is proportional to the slope of the Bode diagram, i.e., of the frequency response of A drawn with logarithmic frequency scale. If this slope is constant at all frequencies and equals $n\,\mathrm{dB/octave}$, then at all frequencies the phase shift is $\sim n15°$.

The weight function $\ln\coth|u/2|$ is charted in Fig. 2.32. Due to its selectiveness, the slope of the Bode diagram in the neighborhood of the frequency ω_c (at which the phase shift is being calculated) contributes much more greatly to B_c than the slope at remote parts of the Bode diagram.

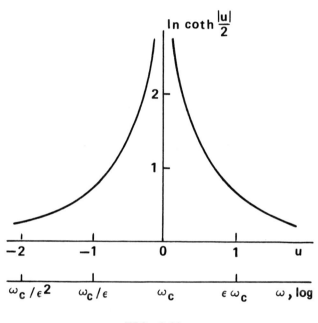

FIG. 2.32

2.3.2 Phase Response Calculation

Piecewise-linear approximation of a Bode diagram, with dA/du kept constant within each of the adjacent intervals, is illustrated in Fig. 2.33. From (2.29), the phase frequency response related to the approximation of the gain response is the sum of integrals, each contributing the phase response component related to a single segment of the frequency response of A, as shown in Fig. 2.33. Bode proved [23] that rather crude approximation of A yields a fairly accurate frequency response for B. For the smooth responses

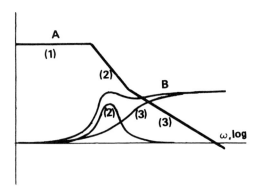

FIG. 2.33

typical for automatic control, the appropriate width of the elementary intervals c is 0.5 to 3 octaves, and the number of the segments is 3 to 5.

Figure 2.34 presents the elementary phase responses: for the ray starting at ω_i and having a slope of $-6n\,\mathrm{dB/octave}$ (dotted line), and for the segments of the gain response with a constant slope of $-6n\,\mathrm{dB/octave}$ over c octaves centered at ω_i, i.e., at $u=1$. Also, a dashed line indicates the phase response related to a single real pole of a transfer function.

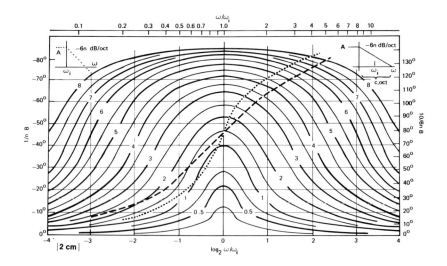

FIG. 2.34

Piecewise-linear approximation of $A(\omega)$ is particularly useful in trial-and-error procedures to find a physically realizable response for $\theta(j\omega)$ that maximizes a certain parameter while complying with a set of heterogeneous constraints (such as weighted maximization of the real component over a given frequency range under a limitation in the form of a prescribed boundary for the frequency hodograph of the function). Starting with some initial response for A, let us say, A', we could then calculate the related response B' and obtain a physically realizable $\theta' = A' + jB'$. Then, further, changing the gain response as seems reasonable, we find the related phase response, *et cetera*. As a rule, the process rapidly converges.

If the desired response $A(u)$ includes sharp changes, such as abrupt banks, its adequate piecewise-linear representation might require more than five segments. The number of segments may be reduced by employing the responses described in the next section.

2.3.3 Piecewise-Constant Real and Imaginary Components

Stability and sensitivity considerations constrain the Nyquist diagram. These limitations are often conveniently expressed in the form of alternate restrictions of the gain and phase, as illustrated in Fig. 2.35. If, under these constraints, the loop gain has to be maximized over the normalized operational frequency band $\omega \leq 1$, then it follows from (2.20) that within this band the loop gain must be flat, and from (2.18) that the phase lag should approach the maximum allowable value. Such a Nyquist diagram closely fits the boundary conditions and is composed of segments alternately parallel to the axes of the Nichols chart. The physically realizable function θ is, therefore,

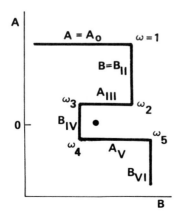

FIG. 2.35

of interest whose components are defined by

$$(A = A_0: \omega < 1), (B = B_{II}: \omega \epsilon [1, \omega_2], \ldots)$$

to be referred to in the following as A_0, B_{II}, A_{III}, *et cetera.**

The simplest, two-segment function θ_l is defined by

$$\text{Re } \theta_l = A_0: \omega \leq 1; \text{ Im } \theta_l = B_{II}: \omega \geq 1$$

This function conformally maps the right half-plane of s onto the interior of a sector on the $\exp \theta_l$-plane, as shown in Fig. 2.36(d) for the case $B_{II} > 0$, and in Fig. 2.36(e) for $B_{II} < 0$. If $B_{II} = \pi/2$, the function $\exp \theta_l$ is positive-real, i.e., satisfying the condition:

$$\text{Re } \exp \theta_l(s) > 0 : \text{Re } s > 0$$

The transform can be performed in three steps: employing the function $W = (\omega^2 - 1)^{1/2} - \omega$, as shown in Fig. 2.36(b); multiplying by j, Fig. 2.36(c); and changing the scales in amplitude and angle, so that

$$\exp \theta_l = (j\omega_1)^{2B_{II}/\pi} \exp A_0 \qquad (2.31)$$

Hence, we have

$$\theta_l = A_0 + \frac{2}{\pi} B_{II} \ln [(1 - \omega^2)^{1/2} + j\omega] = A_0 + \frac{2B_{II}}{\pi} \text{arc cosh } \omega \qquad (2.32)$$

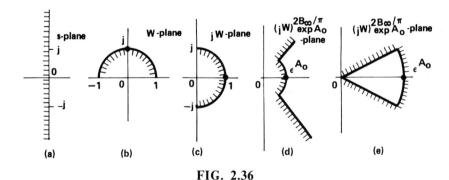

FIG. 2.36

---

*Note in passing that A_0 equals the Laurant series coefficient A_0; the value of A at the last section, if at this section A is constant, equals the Laurant coefficient A_∞.

Note that if $B_{11} = -\pi$, and $A_0 = 0$, expression (2.32) presents the characteristic transmission function of the k-type lowpass filter [23,60]. Hence, the impedance

$$Z_l = \exp \theta_l = q/[j\omega + (1-\omega^2)^{1/2}] \qquad (2.33)$$

where $q = \exp A_0$, is positive-real and realizable as the parallel connection of a capacitance $1/q$ and a two-pole with the impedance equal to the image impedance of Π-end, k-type lowpass filter with the nominal wave impedance $q\Omega$, as shown in Fig. 2.38. The impedance function (2.33) has a unique property: the modulus of Z_l is constant over the band $[0, 1]$ and is larger than that of any positive-real function having the same high-frequency asymptote. Such a two-pole used as the load for a FET amplifier stage, with the parallel stray capacitance $2q$, renders the maximum available gain over the bandwidth $\omega \leq 1$.

From (2.33),

$$\omega = (q/jZ_l + jZ_l/q)/2 \qquad (2.34)$$

For the inverse boundary conditions B_1 and A_∞, inverting the variable gives

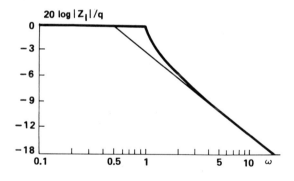

FIG. 2.37

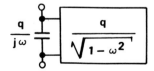

FIG. 2.38

$$\theta(j\omega) = \theta_l(1/j\omega) \tag{2.35}$$

The three-segment function may belong to one of two types. The first type is the function $\exp\theta$ defined by the boundary conditions B_I, A_II, and B_III. Particularly, when $B_\text{I} = B_\text{III}$, the solution follows from the bandpass transform of (2.32). When $B_\text{I} = 0$ and $B_\text{III} = \pi$, we can use the expression for the transfer characteristic coefficient of a filter link with asymmetrical cut-offs [60]. A linear combination of these two partial solutions with a certain constant gives the general solution. This solution is expressed through elementary functions of ω.

The opposite is true for the function with the boundary conditions A_0, B_II, and A_∞. For such a function, $\exp\theta$ maps the right side of the s-plane onto the interior of a sector of a ring, as shown in Fig. 2.39. Using elementary functions, this domain may be mapped onto the interior of a demiellipse, which, in turn, maps onto the right half-plane of s by elliptic integrals. This is why $\exp\theta$ cannot be exactly expressed through elementary functions.

A useful approximation for the case of $\exp A_0 \gg \exp A_\infty$ is described below. Denote $\exp A_0 = q$ and assume $A_\infty = 0$, $B_\text{II} = -\pi/2$. Then $\theta = \ln Z_n$, where

$$Z_n(j\omega) = \frac{q}{Z_l[Z_l(j\omega)/2]} = \frac{Z_l}{2} + \left[1 - \left(\frac{Z_l}{2}\right)^2\right]^{1/2} \tag{2.36}$$

is the solution of the equation:

$$Z_l = Z_n - 1/Z_n \tag{2.37}$$

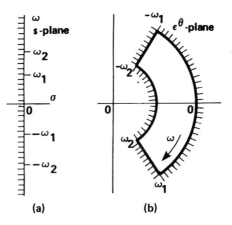

FIG. 2.39

Figure 2.40 shows a series of consequent conformal mappings resulting in (2.36). The function (2.36) maps the right half-plane of s onto the domain shown in Fig. 2.40(c) whose boundary consists of a demicircle, two sections of radial rays, and a demiellipse having the ratio of the main axes:

$$\frac{1 + (1 + 4/q^2)^{1/2}}{1 + (1 - 4/q^2)^{1/2}} \cong 1$$

The approximation of (2.36) conforms exactly to the prescribed boundary for $\omega \geq 1$, and approximately for $\omega < 1$, as illustrated in Fig. 2.41 with $q = 4$, the maximal error in gain being 0.56 dB at $\omega = 1$.

Consider this function in more detail. At frequency ω' of the joint between the first and the second sections, $Z_n = j$; then, from (2.37), $Z_l = -2j$, and, from (2.34),

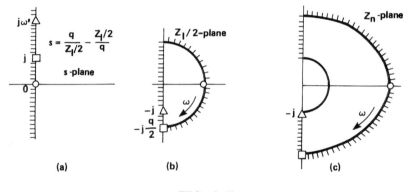

FIG. 2.40

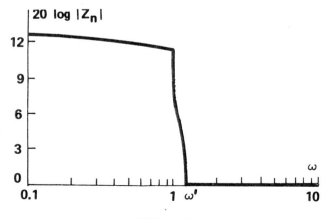

FIG. 2.41

$$\omega' = q/4 + 1/q \tag{2.38}$$

which is plotted in Fig. 2.42.

At $\omega \geq \omega'$, $|Z_n| = 1$; $\arg Z_n = B_{II}$ is plotted in Fig. 2.43.

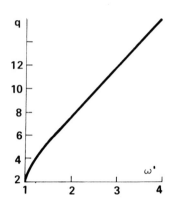

FIG. 2.42

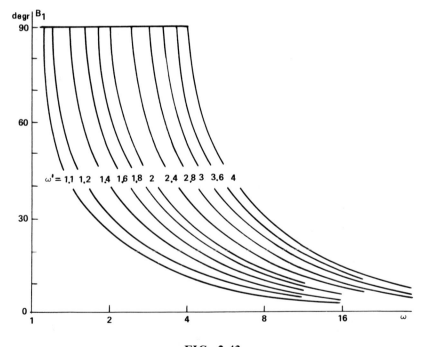

FIG. 2.43

Stability & Frequency Response Constraints

At $\omega \leq 1$, $|Z_l| = q$. Since it is assumed that $q \gg 1$, from (2.36) $|Z_n| \cong q$, so that the relative error is

$$(q - Z_n)/q = 1/2 - (1/4 - 1/q^2)^{1/2} \tag{2.39}$$

Figure 2.44 plots the error *versus* q. When q is large, the error is proportional to q^{-2}.

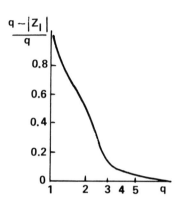

FIG. 2.44

Solution (2.36) is easily generalized to the case of any numerical conditions A_0, B_{II}, A_∞ by substitution of q by $q^{-\pi/2B_{II}}$, raising the obtained expression to the power of $-2B_{II}/\pi$ and multiplying by $\exp A_\infty$.

The transformation (2.36) may be also applied to multisegment functions. For example, applied to functions of the type B_I, A_{II}, $B_{III} = \pi/2$ (which, as noted, can be found analytically), it renders an approximate solution for the problem of four segments: B_I, A_{II}, $B_{III} = \pi/2$, $A_\infty = 0$. This solution meets the boundary conditions exactly at the last two segments, as shown in Fig. 2.45. Frequency inversion gives the gain response plotted in Fig. 2.46.

General solution for the four-segment problem is expressed through elliptic integrals [23]. The general solution for a five or more section problem cannot be expressed through conventional tabulated functions.

A gain response composed of alternating sections of finite and zero slope shown in Fig. 2.47 comprises a rough solution to the multisegment problem, with rounded corners for the locus of the functions on the Nichols plane, as will be clarified later by Fig. 8.16, and lower available gain in the band of interest.

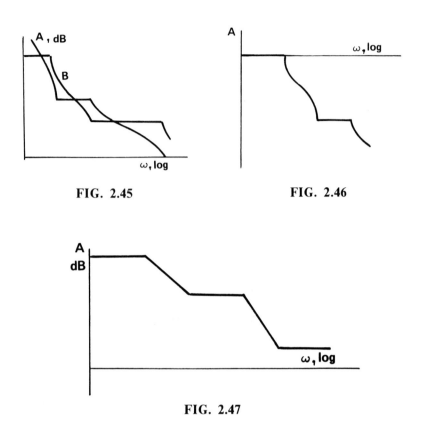

FIG. 2.45

FIG. 2.46

FIG. 2.47

2.3.4 Examples and Exercises

1. Impedance Z of a passive two-pole includes a parallel capacitance C. Find the maximum value for $|Z|$ available over the frequency band $[0, \omega_0]$.

From (2.33), $C = 2q$ and

$$Z = \frac{C/2}{j\omega/\omega_0 + [1 - (\omega/\omega_0)^2]^{1/2}} \tag{2.40}$$

so that $|Z| = 2/\omega_0 C$ for $\omega \leq \omega_0$. With $\omega_0 = 1$, frequency responses for $A = 20 \log|Z|$ and $B = \arg Z$ are plotted in Fig. 2.48.

2. The output of a FET having the transconductance of 0.1 S is shunted by a stray capacitance of 0.9 pF. Find the maximum available flat gain over the bandwidth of 1 GHz.

Using (2.40), the maximum available modulus of the load impedance is 50 Ω. Hence, the maximum value of the amplification is 5.

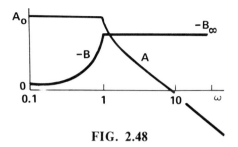

FIG. 2.48

3. Frequency responses for $R = \mathrm{Re}\, Z$ and $X = \mathrm{Im}\, Z$ of function (2.40) are shown in Fig. 2.49. We see that R is 0 outside the working band. The two-pole can, therefore, be realized in Darlington's form [23,83], as the input impedance of a reactive two-port loaded at a resistance $R = 2/\omega_0 C$. The reactive two-port represents a filter of infinite selectivity, as depicted in Fig. 2.50(a,b), for lowpass and bandpass versions, respectively. The attenuation of a realizable filter with a finite number of elements cannot be infinite over a finite bandwidth. Then, R may not equal 0 throughout the stopband. The realizable response for Z shown in Fig. 2.51 by the solid line differs from the idealized response (dotted line).

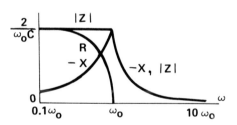

FIG. 2.49

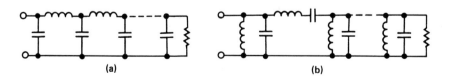

FIG. 2.50

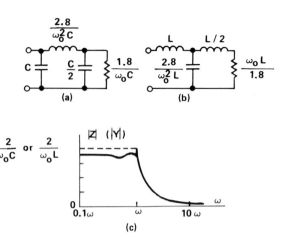

FIG. 2.51

4. Find the maximum available value A_0 of Re θ remaining flat over the frequency band $\omega \leq 1$, if $B = \text{Im}\,\theta \geq -5\pi/6$, and at a certain given frequency $\omega_b \gg 1$, $A(\omega_b) = 0$.

The frequency response for θ is readily obtained by multiplying A and B related to the function (2.40) by 5/3. That is, the frequency response for A is similar to that shown in Fig. 2.37, but has the asymptotic slope of $-10\,\text{dB/octave}$. Therefore,

$$A_0 = 10 \log_2 \omega_b - 10$$

5. Compute the response of B related to the response of A plotted in Fig. 2.52.

First, the given response for A is approximated in piecewise fashion with reasonable accuracy. Then, using the charts of Fig. 2.33, the phase shift components are plotted, generated by each of the segments of constant slope. Summing the phase components gives the desired response for B. However, to make the calculations easier, we shall compose the approximation for the response of A with only three elementary responses: with a slope $-6\,\text{dB/octave}$ over the entire ω-axis, which contributes $-90°$ to the phase shift response at all frequencies; a ray with a slope of $-12\,\text{dB/octave}$ starting at frequency ω_3; and the segment having a slope of $-6\,\text{dB/octave}$ over the interval $[\omega_1, \omega_2]$. Summing the elementary phase shift responses gives the required response $B = -\pi/2 + B_r + B_s$.

6. Approximate the solution to the three-segment problem of the type A_0, $B_1 = -\pi/2$, $A_\infty = 0$ through composition of the solutions to the two-segment problems.

Stability & Frequency Response Constraints

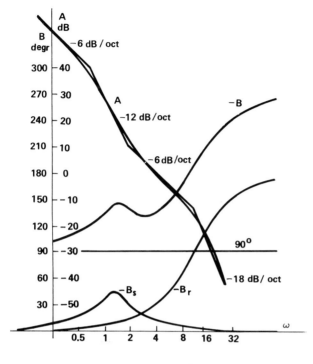

FIG. 2.52

By frequency inversion, using the new variable $\omega'/j\omega$, we obtain from (2.33):

$$\frac{q}{Z_l(\omega'/j\omega)} = \frac{\omega'}{j\omega} + [1 - (\omega'/\omega)^2]^{1/2} \tag{2.41}$$

with the frequency response charted in Fig. 2.53.

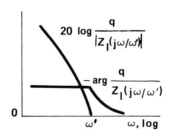

FIG. 2.53

Next, by setting $\omega' = q/4$, we provide the high-frequency asymptote of $Z_l(j\omega)$ equal to the low-frequency asymptote $q/2j\omega$ of expression (2.41). Subtracting this asymptote $q/2j\omega$ from the sum of expressions (2.33) and (2.41) produces the function:

$$\exp\theta = Z_n = Z_l + \frac{q}{Z_l(\omega'/j\omega)} - \frac{q}{2j\omega} \qquad (2.42)$$

which is illustrated in Fig. 2.54 (a,b) for $q=8$. This approximation is less accurate than that of (2.36); in the worst case, at $\omega = 1$, the error is 2 dB (instead of only 0.04 dB).

Using the same method for the inverse values, another approximation is $1/Z_l + [Z_l(\omega'/j\omega)]/q - 2j\omega/q$.

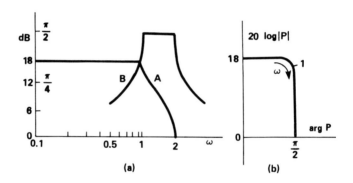

FIG. 2.54

2.4 FEEDBACK MAXIMIZATION

2.4.1 Bode Optimal Cut-Off

The Bode feedback maximization problem is stated as follows:

$$\left. \begin{array}{l} \min |T| = \max : \omega \leq 1 \\ |T| \gg 1 : \omega \leq 1 \\ |T| = (\omega_c/\omega)^n \, 10^{x/20} : \qquad \omega > \omega_c \\ |\arg T| \leq (1-y)\pi \\ |B_n| \cong |B_{nc}\, \omega/\omega_c| \leq 1 \text{ rad} \end{array} \right\} : \omega < \omega_c \qquad (2.43)$$

Stability & Frequency Response Constraints

As we can see, $|T| \cong |F|$ is being maximized over the normalized operational band $\omega \leq 1$. The frequency ω_c is defined as the frequency at which $20 \log |T| = -x \, \text{dB}$. Beyond ω_c, $20 \log |T|$ decreases with the constant slope (asymptotic slope) of $-6n \, \text{dB/octave}$. At ω_c, the value of the nonminimal phase shift (NPS) B_{nc} of the loop transmission is assumed to be known and less than 1 rad. Then, as will be proved in Sec. 2.5.1, for $\omega < \omega_c$, the loop NPS B_n is approximately proportional to ω. The system is required to remain stable with the phase stability margin no less than $y\pi$, when the loop gain is increased at all frequencies by $x \, \text{dB}$, or decreased by any value.

The approximate solution to the problem known as the *Bode optimal cut-off* is charted in Fig. 2.55 (a,b,c). The Bode diagram for T is composed of three elementary responses: the solution to the two-segment problem (recall Sec. 2.3.3) with the asymptotic slope of $-(1-y)12 \, \text{dB/octave}$ and the related

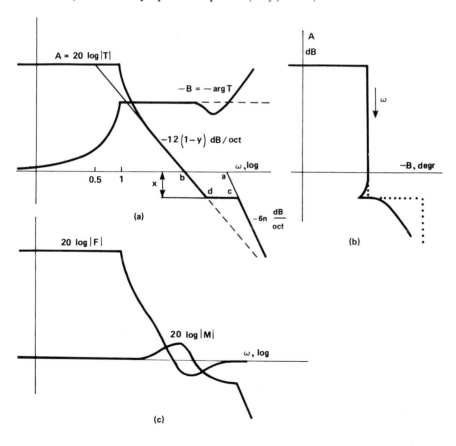

FIG. 2.55

phase shift of $-(1-y)\pi$ shown by the dashed lines, the step at the level of $-x\,\mathrm{dB}$ between the frequencies ω_d and ω_c, and the ray with the slope of $-6n\,\mathrm{dB/octave}$ starting at ω_c.

At lower frequencies, arg T is required to retain the value of the solution for the two-segment problem. Hence, at lower frequencies, the phase shift contribution from the ray with the slope of $-6n\,\mathrm{dB/octave}$, which starts at ω_c, together with B_n, should equivalently replace the phase contribution of the discarded ray shown by the dashed line, which has the slope of $-12(1-y)\,\mathrm{dB/octave}$ and starts at ω_d; i.e.,

$$-2n\omega/\pi\,\omega_c - |B_{nc}|\,\omega/\omega_c = -4(1-y)\omega/\pi\,\omega_d$$

Therefore, we have

$$\omega_d = 2(1-y)\,\omega_c/(n + |B_{nc}|\,\pi/2) \tag{2.44}$$

Hence, according to the chart of Fig. 2.55, the feedback in dB in the working band is given by

$$A_0 = 20\log|T| = 12(1-y)(1 + \log_2\omega_d) - x$$

or, further,

$$\begin{aligned}A_0 = 12(1-y)\,[2 + \log_2(1-y) - \log_2(n + |B_{nc}|\,\pi/2)] - x \\ + 12(1-y)\log_2\omega_c\end{aligned} \tag{2.45}$$

In the usual case of $y = 1/6$,

$$A_0 = 17 - x - 10\log_2(n + |B_{nc}|\,\pi/2) + 10\log_2\omega_c \tag{2.46}$$

which is presented in graphic form in Fig. 2.56.

The value of A_0 in (2.56) yields typically only 2 to 3 dB to the ideal solution for the system (2.43) shown in Fig. 2.55(b) by the dotted line. Better approximations of this solution (as demonstrated by Bode [23], or achievable as a solution to the problem of four segments related to Fig. 2.46) seem impractical, since their accurate implementation requires increasing the order of the function $T(s)$ beyond that which is reasonable.

The bandpass transform [23,83] applied to the lowpass response of Fig. 2.55(a) yields the optimal cut-off for a bandpass system, as illustrated in Fig. 2.57(a,b). The transform retains the bandwidth of the available feedback regardless of the relative bandwidth. The smaller the relative bandwidth, the steeper is the slope of the cut-off. When the relative bandwidth is fairly wide, e.g., more than 2 octaves, the steepness of the low-frequency

Stability & Frequency Response Constraints

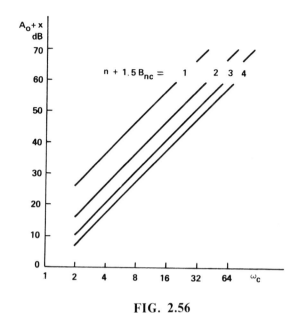

FIG. 2.56

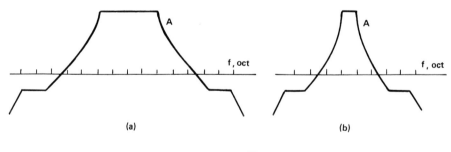

FIG. 2.57

slope of the response has almost no effect on the available feedback. Practical cut-off implementation in a multioctave bandpass system is not symmetrical, with shallower gain slope and larger phase stability margin at lower frequencies.

Consider next the physical constraints on ω_c (and, therefore, on $\omega_b \cong \omega_c/2$).

1. Assume that the wideband feedback amplifier to be designed contains n stages, where the gain of each decays at higher frequencies with the slope of $-6\,\text{dB}/\text{octave}$. Denoting as ω_{Ti} the frequency at which the gain on the ith stage is $0\,\text{dB}$, the total amplifier gain becomes $0\,\text{dB}$ at the logarithmic average

frequency:

$$\omega_{Tav} = \left(\prod_{i=1}^{n} \omega_{Ti}\right)^{1/n}$$

At this frequency, therefore, the loop gain equals with opposite sign the attenuation of the passive links in the loop, i.e.,

$$-20 \log |T(j\omega_{Tav})| = A_T$$

as shown in Fig. 2.58.

In actual design $A_T \epsilon$ [8,20] dB and, as will be explained in Sec. 2.4.3, remains constant within 1 to 2 octaves beyond ω_c. Then ω_c is found from Fig. 2.58 as

$$\log_2 \omega_c = \log_2 \omega_{Tav} - (A_T - x)/6n$$

i.e.,

$$\omega_c = \omega_{Tav} \, 2^{(x - A_T)/6n} \qquad (2.47)$$

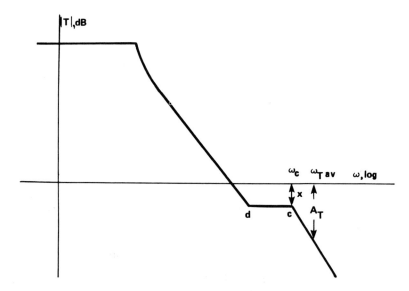

FIG. 2.58

Stability & Frequency Response Constraints

Increasing n reduces the ratio ω_d/ω_c of (2.44), and in the usual case of $A_T > x$ increases ω_c by virtue of (2.47). These two effects combined produce a broad maximum in the dependences $\omega_d(n)$ and $A_0(n)$ in the region of n from 2 to 3. This is why wideband feedback amplifiers usually contain 2 to 4 stages of amplification.

2. Figure 2.59 illustrates the typical lowpass characteristics of the object of regulation Q and of T in the control system of Fig. 1.32. The noise effect at the input of the nonlinear link given by (1.54) does not depend on $\mu\mu_1$ at lower frequencies, where $|T| \gg 1$. At higher frequencies beyond ω_b, reducing $|T|$ *versus* ω (as quickly as the requirement allows for preserving the stability margins) benefits the noise reduction. With this, the noise manifests itself only at the frequencies within 2 to 4 octaves above ω_b, where the denominator in (1.54) is small and the numerator is large.

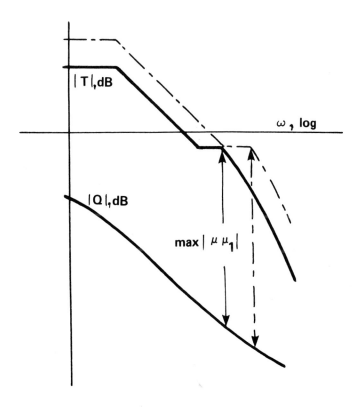

FIG. 2.59

It can be seen from Fig. 2.59 that the increase of the feedback (dashed-dotted line) is attained at the price of increasing $|\mu\mu_1|$, i.e., at the price of greater noise effect at the input to the nonlinear link in Fig. 1.32. Hence, the permissible upper limit of the noise effect restricts the available feedback in the operational band by constraining ω_b.

3. Including a sample data link in the feedback loop does not affect the stability condition when the sampling frequency $\omega_s \gg \omega_c$, practically, when $\omega_s > 3\omega_c$. When ω_s is not that high, the phase lag of the time-invariable links of the loop must be reduced by up to $\pi/3$ in the frequency range of $\omega_b/3$ to $3\omega_b$. This reduces the maximum admissible slope of the Bode diagram and the available feedback.

With a digital processor in the loop, the insufficient computing power reduces the available feedback in two mutually related ways. First, the available feedback is reduced because of smaller than desired ω_s, as described above. Second, the limited volume of computation to be performed during the sample period restricts the allowable order of the loop compensator, thus increasing the approximation error, and therefore requiring larger average stability margin. Hence, in accordance with the phase integral, the available feedback is reduced.

4. In some frequency range, the ignorance of the loop transmission function might be excessively large (for example, due to resonance modes in flexible mechanical structures). Therefore, ω_c needs to be kept below this range.

5. When the plant is of a lowpass kind, the tolerances of the plant transmission function typically increase with frequency. It has been shown in the literature [20,54] that the tolerances of the filter attenuation are proportional to the slope of attenuation-frequency response (in dB/octave), as illustrated by limiting curves 1 and 2 in Fig. 2.60. Such tolerances interfere with provision of sufficient stability margins at higher frequencies for the curve 1 having a bigger slope. Because of these tolerances, ω_c and the available feedback are limited.

When the tolerances exceed 15 dB and the sensitivity widely varies over this range of parameter variations, the design methods developed by Horowitz et al. [69,72] can be employed.

2.4.2 More Cut-Offs

Three sets of objectives are typically used for feedback systems design:

Minimization of sensitivity or nonlinear distortions over the normalized band of operation $\omega \leq 1$, either flatly or weighted, with prescribed frequency response. Frequency responses like those plotted in Fig. 2.61 are implied. These responses satisfy the limitation (2.20).

Stability & Frequency Response Constraints

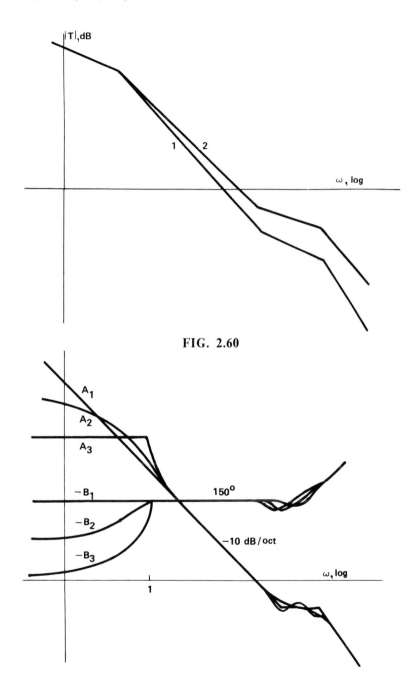

FIG. 2.60

FIG. 2.61

If maximization of $|S_m|$ is the principal requirement, the frequency response of $T(j\omega)$ may be smooth, as shown in Fig. 2.62, allowing for lower-order realization of the compensator. With smaller $|T|$ but also smaller $|\arg T|$ at $\omega = 1$, the function $|S_m|$ almost retains the value related to the Bode optimal cut-off. Certain responses for T increase $|S_m|$ over the optimal Bode cut-off, like that shown in Fig. 2.63.

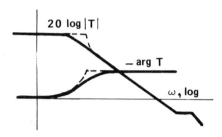

FIG. 2.62

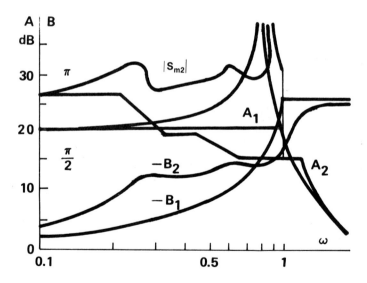

FIG. 2.63

Good transient response, static and dynamic accuracy of tracking. Step function is commonly employed as the signal for testing the performance of linear systems. The rise time t_r, the overshoot, and the error after the prescribed settling time t_s are the main parameters for the output transient response evaluation, as shown in Fig. 2.64. For tracking systems, the objectives can be translated into the frequency domain for M as in the following.

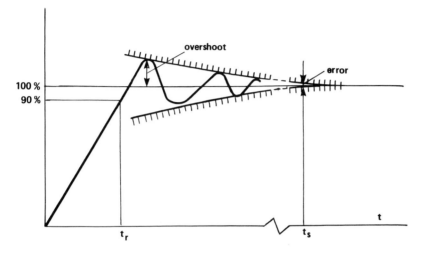

FIG. 2.64

The rise time and the overshoot belong to the range of smaller time values. They relate to the behavior of the frequency response of M at the upper edge of the frequency range where $|M| = 1$, i.e., as is seen in Fig. 2.55, at frequencies near ω_b. The frequency response of M shown in Fig. 2.55 fails to provide good transient response.

A linear corrector installed at the input to a system seemingly can make the total response close to the ideal (reminding us of the responses of Butterworth or linear phase filters). However, the plant gain variations change T and ω_b, and greatly change the system closed-loop gain in the neighborhood of ω_b, since here the feedback is positive. The input corrector compensating for these variations would be required to be time-variable, which is not practical. Hence, the frequency response of the closed-loop feedback system must inherently preserve the desired smooth shape over the entire range of possible ω_b. For this, the phase stability margin ought to be large enough (over 45° or 60°) in the relatively wide band around average ω_b, and the loop gain response must be similar to that shown in Fig. 2.65.

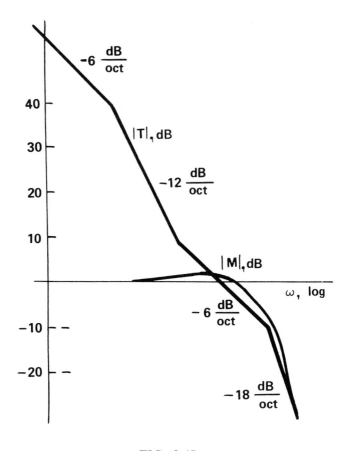

FIG. 2.65

Adding or subtracting constant gain tolerances to such loop gain response still results in smooth response of M, thus guaranteeing acceptable tracking transient response.

Because the slope of the feedback frequency response within the interval 1 to ω_b is typically -8 to $-10\,\mathrm{dB/octave}$, i.e., $|F|$ decreases with frequency as $\sim \omega^{-1.5}$, the envelope of the time-response error narrows as $t^{-1.5}$ over the time interval $1/\omega_b$ to 1.

Minimal quadratic index of the output time response. Due to the simple mathematical formulation of the problem and the existence of elaborate mathematical methods for solving such problems, this approach seems attractive. It has found wide application in optimal filtering, prediction, and smoothing in linear systems. However, when applied to closed-loop control problems, it frequently leads to unacceptable transient response, insufficient feedback around the plant, a wind-up (see Sec. 4.2), and far from optimal solutions for multiple-loop systems. A detailed criticism of such approach deficiencies is found in the literature (Horowitz et al. [69,71]). In its present form, the approach cannot solve the problem of feedback maximization, and therefore is considered to be beyond the scope of this book.

As mentioned, and which will be detailed in Chapter 3, the requirements of (2.43) are tailored to guarantee asymptotic global stability of the system with a single, memoryless, nonlinear link in the loop. If, however, the system is furnished with an additional dynamic nonlinear link, the constrains of (2.43) could be loosened and the available feedback increased by application of the Nyquist diagram shown in Fig. 2.66 [24].

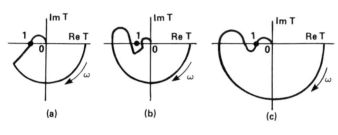

FIG. 2.66

The type a response uses zero amplitude stability margin, thus increasing A_0 in (2.45) by x dB. In such a system, a periodical self-oscillation takes place at the frequency ω_b. The amplitude of the oscillation must be kept small, as will be discussed in Sec. 5.1.3.

The type b Nyquist diagram does not encircle the critical point, and the system is free from small-signal oscillation. This type of cut-off can be specified as follows:

$$\left. \begin{array}{r} \min |T| = \max : \omega \leq 1 \\ |T| \gg 1 : \omega \leq 1 \\ T = (\omega_c/\omega)^n \, 10^{x/20} : \omega > \omega_c \\ 20 \lg T \notin [-x, x_1] : |\arg T| < (1-y)\pi \\ B_n \cong \omega \, B_{nc}/\omega_c \\ -\arg T \leq -B_u(A) : A > x_1 \end{array} \right\} \quad (2.49)$$

where x_1 and x represent the upper and lower amplitude stability margins in dB, and the boundary curve $B_u(A)$ restricts the permissible location of the Nyquist diagram.

This specification differs from (2.43) by the fact that, at frequencies where $A > x_1$, the phase shift $B = \arg T$ is restricted by a certain given boundary curve B_u, but not by a constant phase stability margin. The curve B_u follows from the properties of a chosen nonlinear dynamic compensator, which will be discussed in Chapter 5. Normally, B_u decreases with frequency as illustrated in Figs. 2.67 and 2.68. In this example, the available feedback is 10 dB larger than that in a system with the Bode optimal cut-off. The global stability of the system is provided by the nonlinear dynamic compensator.

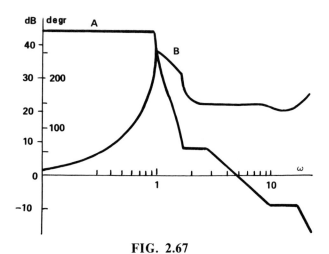

FIG. 2.67

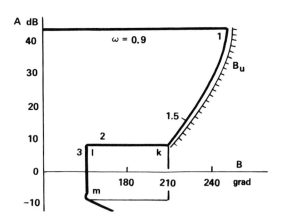

FIG. 2.68

A linear combination of the Bode optimal cut-off and the characteristic transmission function of a lowpass lossy filter of type m gives a rough solution to the system (2.49), as shown in Fig. 2.69 [65, 151]. Better accuracy could be achieved using a linear combination of responses, which are the solutions to three- and four-segment problems described in Sec. 2.3.3, as employed in Sec. 8.2. More methods of finding the response are presented in Sec. 5.3 and Sec. 6.4.2.

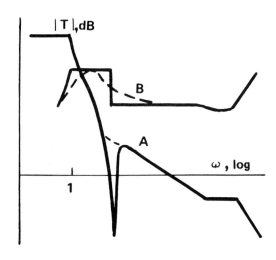

FIG. 2.69

The essential features of the simplified response are the two Bode steps shown in Fig. 2.70, the width of the lower one calculated with (2.44), and that of the upper through a similarly developed formula, which uses the slope of the Bode diagram to the left of the upper step instead of the asymptotic slope.

Due to larger feedback, the time-response error in such systems reduces during the same settling time to a lower value *versus* that of a system complying with the constraints of (2.43).

The system with the Nyquist diagram of Fig. 2.64(c) combines the features of those with the diagrams of 2.64(a) and 2.64(b).

If needed, the shape of the cut-off of T can be smoothly changed using variable equalizers. For example, the L-type corrector of Fig. 1.31 with $Z_1 = Z_l$, $Z_2 = 1/Z_n$ (recall Sec. 2.3.3) yields $Q = Z_n$ [109]. The T-type corrector provides realization of more sophisticated Q functions.

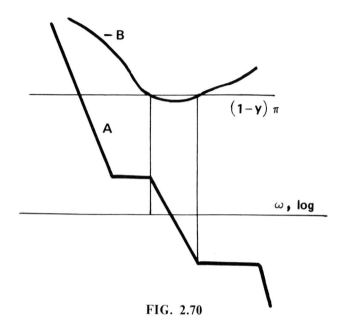

FIG. 2.70

2.4.3 Asymptotic Losses, High-Frequency Bypass, Loop Gain Correction in Feedback Amplifiers

It was shown in Sec. 2.4.1 that in order to increase the maximum available feedback, the asymptotic losses A_T should be kept low. Therefore, the β_0 part and the input and output six-poles must include high-frequency bypasses, like those formed by capacitors in the block-diagram of Fig. 2.71. The capacitance of these capacitors, however, cannot exceed a certain limit, beyond which these elements interfere with the desired performance of the subcircuits in the working band. The maximum admissible values of the capacitances can be evaluated with the help of integrals (2.12) and (2.15).

The admissible value of bypass capacitances might be increased if the subcircuit is transformed so as to lower the impedances faced by the bypass capacitors. Such a transformation of the β_0 circuit is illustrated in Fig. 2.72. The transmission coefficients of the circuits (a) and (b) is the same in the working band (through the resistive network), but, due to lower parallel resistances, the equivalent capacitance of the bypass in circuit (b) is four times greater than that of circuit (a). Another configuration yielding larger bypass capacitance is circuit (c).

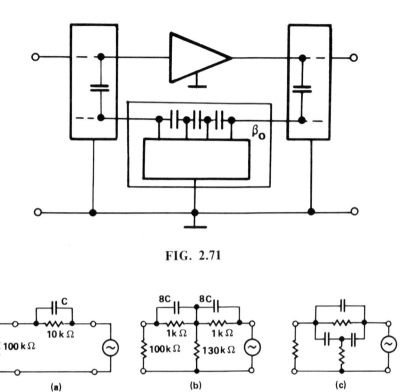

FIG. 2.71

FIG. 2.72

Figure 2.73 shows the equivalent circuit for the frequencies close to ω_c. The two-poles Z_a and Z_b, having low impedance in the working band, and high impedance near the frequency ω_c, separate stray parallel capacitances of the input and output terminals from the feedback loop to keep A_T smaller.

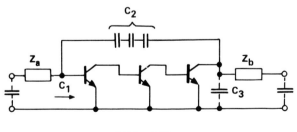

FIG. 2.73

The only capacitances shunting the loop, therefore, are the stray capacitance of the amplifier output C_3 and the input capacitance C_1 of the first transistor, which together with the bypass capacitance C_2 form a capacitive signal divider. The attenuation of the divider:

$$A_T = 20 \log \frac{I_{c3}}{I_{b1}} = 20 \log \frac{C_3}{C_3 + C_1 C_2 / (C_1 + C_2)}$$

is independent on the frequency.

In configurations like that of Fig. 2.71(c), the subcircuit with the bypasses is no longer of the ladder type (whose transmission is always MP). An improperly designed bypass renders zeros of the loop transmission function approaching the $j\omega$-axis, or even crossing it and migrating in the right half-plane. Such zeros cause collapses in the loop gain response, the right-hand zeros also introducing NPS (pure delay) in the loop. These topics will be discussed in Sec. 2.7.3.

After the bypasses have been provided, we must synthesize the linear correctors (compensators, as they are called in automatic control) for leveling out the excess loop gain. These compensators are often installed in the interstage circuits, or in the feedback paths of local feedback loops. Many instructive examples are included in Bode [23], and additional examples can be found in other references [16, 18, 59, 79, 103].

2.4.4 Prediction and Feedback Maximization

The problem of prediction as defined in Fig. 2.74 is that of real-time evaluation of the signal $u_1(t)$ that, if being applied to the input of the given physical linear link $Q(s)$, causes its output to equal the prescribed function $u_2(t)$. If Q were precisely known, no noise sources existed, and the link $1/Q$ were realizable, its output response $u_2(t)$ would exactly restore the signal $u_1(t)$ to be estimated. However, the unavoidable NPS in Q and incomplete knowledge of Q (which could vary with time and be nonlinear) admit only approximation of the function $1/Q$ by a certain realizable function $1/Q_a$. Also, of course, the noise of the amplifiers and sensors contribute to the error.

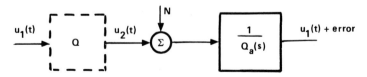

FIG. 2.74

The Fourier transform of the output in response to the input $U_2(j\omega)$ is seen as

$$U_1 + [N/Q_a + U_2(1/Q_a - 1/Q)] \qquad (2.50)$$

The two terms in square brackets represent two components of the error, the first due to the noise N, and the second due to discrepancy between Q and Q_a.

To understand the trade-off between minimization of these error components, let us consider the usual case of Q possessing lowpass properties, thus, assuming $1/|Q|$ increasing infinitely with frequency. To limit the effect of the noise, $1/|Q_a|$ must, therefore, be bounded at high frequencies, even though this increases the second component of the error. The available trade-off depends on the signal statistics and the applied figure of merit.

Another way of estimating U_1 is by looping the link Q with feedback, as shown in Fig. 2.75. The output Fourier transform is

$$U_1 + [MN/Q + U_2/(QF)]$$

where $M = T/F$, and $T = \mu Q$.

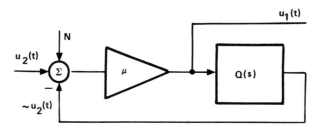

FIG. 2.75

In the frequency range of large feedback, $|M| \cong 1$ and the noise effects in (2.50) and (2.51) are almost equal. At higher frequencies, beyond ω_b, M degenerates into T, and the noise effect becomes $N|\mu|$, thus dying down with frequency. If $Q/M = Q_a$, the noise effects in both systems are exactly equal.

The advantage of the system of Fig. 2.75 over that of Fig. 2.74 is in reduction of the second error component by the feedback. The smaller are the NPS in Q, the ignorance in Q at higher frequencies, and the noise N, then the greater is the available feedback or the width of the frequency range where the available feedback is large. The broader is this range, then the more quickly varying signals can be predicted. The larger the feedback, the higher is the accuracy of prediction. Therefore, enhancing the feedback by exploitation of Nyquist-stable systems improves the accuracy of prediction over that available in a system with linear loop compensators.

2.4.5 Negative Resistance Sensitivity

An active two-pole is often employed to compensate the losses of a resonator or a transmission line. In this case, precision realization of its impedance Z is of prime importance.

The feedback system of Fig. 1.4 may be used as such an active two-pole. The larger the feedback, the smaller is the modulus of sensitivity S_z of the impedance to the amplifier coefficient of amplification. The minimum available $|S_z|$ depends on the range of load impedances to which the system must be coupled and remain stable.

Stability of the system may be studied with Nyquist or Bode diagrams for the amplifier. The parts of the diagram out of the operational frequency band can be corrected by an additional two-port embedded between the active two-pole and the load. Within the operational band, the attenuation of the two-port is required to be small and the amplifier return ratio depends on the loading impedance.

We illustrate this problem with the study of a two-pole exhibiting negative resistance $Z = \text{Re } Z = -r < 0$ over the operational band $\omega \leq 1$. The two-pole is loaded into a positive resistance R_L. Examine three ranges of values for R_L.

Suppose the system must be stable when $R_L > r$. Therefore, $Z(s)$ is allowed to have zeros, but not poles, in the right half-plane of s. Since the circuit must be stable with the load impedance infinite, the frequency hodograph of $T(\infty)$ may not encompass -1. Hence, the maximum $|T(\infty)|$ available in the working band is attained if $T(\infty)$ has the frequency response of the optimal cut-off $T_B(j\omega)$. With Z_0 given, $T(0)$ is found from Blackman's formula. The sensitivity S_z is then found from (1.39).

The requirement for the circuit to be stable when $R_L < r$, is dual to the already considered problem. To minimize the sensitivity, $T(0)$ has to be realized as the optimal cut-off.

If the circuit is required to be stable with any R_L within a certain finite interval not including 0, the function $Z(s)$ is allowed to have both zeros and poles on the right. Suppose further that the element values of the feedback circuit are all real within the working band, and $Z = R_0$ is real and positive. Then, it can be seen from Blackman's formula that $T(0)$ differs from $T(\infty)$ by only a real multiplier. From (1.22), the frequency response for the return ratio:

$$T(R_L) = \frac{T(0) R_0 + T(\infty) R_L}{R_0 + R_L} \qquad (2.52)$$

therefore, possesses the same shape as $T(0)$ and $T(\infty)$. For the feedback to be maximum, we assume that these responses differ from $T_B(j\omega)$ by only real multipliers. Then, the condition for the locus of $T(R_L, j\omega)$ not to encompass the critical point -1 can be expressed as

$$-1 < T(R_L) < T_B : \omega = 0 \qquad (2.53)$$

Using (2.52) and Blackman's formula $-r = R_0[T(0)+1]/[T(\infty)+1]$, the inequality (2.53) is transformed into

$$r < R_L < -R_0 \frac{T(0) - T_B}{T(\infty) - T_B} : \omega = 0 \qquad (2.54)$$

At zero frequency, either $T(0)$ or $T(\infty)$ is negative. If $T(0)$ is negative, then $R_L > -r$. Otherwise, $R_L < -r$.

When the range of admissible R_L is narrow, both $T(0)$ and $|T(\infty)|$ may be considerably larger than $|T_B|$, thus reducing $|S_z|$ given by (1.39). With $|T(0)| \gg |T_B|$ and $|T(\infty)| \gg |T_B|$, the relative width of the admissible interval of load resistances is

$$\frac{R_L + r}{-r} = T\left(\frac{1}{T(0)} + \frac{1}{T(\infty)}\right) : \omega = 0 \qquad (2.55)$$

2.4.6 Examples and Exercises

1. Figure 2.76 displays a fiber-optic receiver of *transimpedance* type. The capacitance C_1 represents the sum of the input capacitance of the photo PIN-diode, of the input of the FET, and the stray capacitance. Large parallel feedback makes the input impedance of the amplifier several times lower than the impedance $1/j\omega C_1$. Over the bandwidth where this inequality sustains, the effect of C_1 on the output signal is negligible, and the frequency response of the ratio of the output voltage to the current source (transimpedance) equals the impedance of the feedback path two-pole.

It is worth increasing the feedback path resistor R_1 to reduce the thermal noise current $\overline{I}_n = (4kT\Delta f/R_1)^{1/2}$ applied to the amplifier input. Maximum value of R_1 is limited by the requirement that the impedance of parallel connection of R and the bypass capacitance C_2 must not change more than by 1 dB within the operational frequency band, in order to keep the frequency

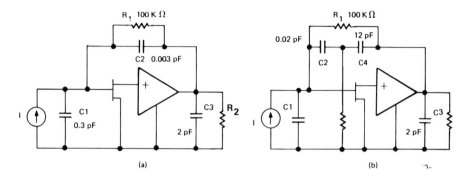

FIG. 2.76

response of the transimpedance flat. The capacitance C_2 must be sufficient to guarantee large feedback over the operational frequency band.

The transistors have unit gain frequency $f_{Tav} = 7.4$ GHz; $R_1 = 100\,k\Omega$, $R_2 = 200\,\Omega$, $C_1 = 0.3$ pF, and $C_3 = 2$ pF. Calculate what feedback is realizable over the bandwidth of 100 MHz.

The bypass capacitance C_2 is readily found to be 0.003 pF from the condition that the modulus of its impedance must be at least three times greater than R_1 at 100 MHz. The asymptotic losses are

$$A_T = 20 \log C_3/C_2 = 56 \text{ dB}$$

With three stages of amplification, the asymptotic slope of the loop gain is -18 dB/octave. Then, from (2.47),

$$f_c = f_{Tav} 2^{(10-56)/18} = 1.2 \text{ GHz}$$

Assuming $B_{nc} = 1$ rad (as seen in Fig. 2.56, the value of B_{nc} is not critical), from (2.44),

$$f_d = 0.44 \text{ GHz}$$

Next, we plot the Bode diagram in Fig. 2.77 that seems optimal for the system in point. The feedback at the frequency of 100 MHz is 12 dB. This value is theoretical, realizable when the frequency response for T is implemented exactly as desired. Considering the deviations of a practical low-order implementation from this curve, the bandwidth of the large feedback reduces to 70 MHz.

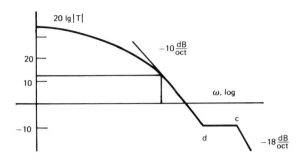

FIG. 2.77

If the frequency band of operation needs to be increased, the value of A_T must be reduced by increasing C_2. With increased C_2, maintaining the flat gain-frequency response will require reducing R_1 and the signal-to-noise ratio.

A better solution is revealed by Fig. 2.76(b). The improved feedback path consists of a two-link highpass filter having the same attenuation in the working band as that in the feedback path of Fig. 2.76(a). Due to better filter selectivity, A_T is appreciably reduced. The capacitance $C_4 = 0.2\,\text{pF}$ is chosen to be much smaller than C_3 so as not to shunt the output. The capacitance $C_2 = 0.02\,\text{pF}$ being 15 times smaller than C_1 causes little impairment of the signal-to-noise ratio. For this circuit, $f_c = 2.25\,\text{GHz}$, $f_d = 0.82\,\text{GHz}$, and therefore the available theoretical bandwidth of 12 dB feedback extends up to 180 MHz; or, for practical purposes, to 125 MHz.

2. The simplified circuit diagram of a fiber-optic receiver is displayed in Fig. 2.78. A PIN-photodiode is coupled to the input of a FET stage followed by a two-stage bipolar transistor amplifier with the feedback parallel at its input and output. The amplifier transimpedance is flat up to 0.3 GHz and then increases with the slope of 6 dB/octave because of the shunting RC

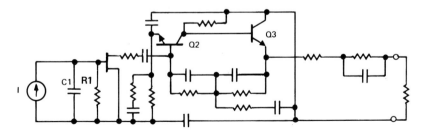

FIG. 2.78

branch in the feedback path. The noise contributed by the resistor with large resistance R_1 is negligible. The stray capacitance C_1 integrates the input current from the equivalent current source. The equalizers at the input and output of the amplifier make the total load-source response flat up to 0.3 GHz. The response also remains flat beyond 0.3 GHz, up to 0.7 GHz, due to increasing gain of the amplifier compensating the input integrator downslope. The high-frequency bypass is implemented as a two-link RC-filter to make A_T low. The emitter local feedback in Q_2 reduces the common-loop gain at lower frequencies to reduce the slope of the Bode diagram and provide the circuit stability. Note that the employed element connection shortens the physical length of the feedback path at higher frequencies from the output terminals of the amplifier (collector, emitter of Q_3) to the input of the amplifier (emitter, basis of Q_2).

3. The importance of accurate approximation of the desired frequency responses is illustrated by the improvements made in hybrid, thin-film, push-pull linear amplifiers, a simplified schematic of which is shown in Figs. 2.79 (before modification) and 2.80 (after modification). Note that the added circuit elements are printed with thin-film technology at no extra cost.

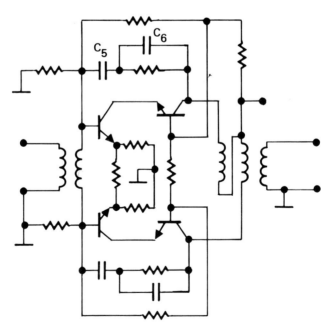

FIG. 2.79

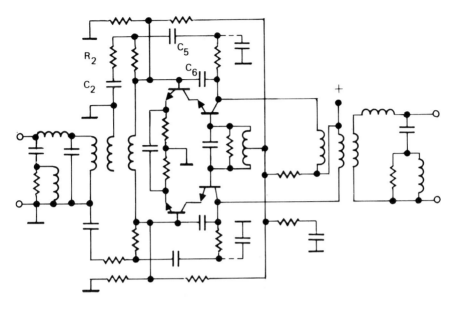

FIG. 2.80

First, consider the changes in the parallel feedback path. Placing the blocking capacitor C_5 between the two parts of the formerly single feedback resistor renders twofold improvement: the parasitic-to-ground capacitance of the physical blocking capacitor does not now shunt the signals at the asymptotically high frequency, i.e., A_T is reduced, and this capacitance together with the added branch of the printed elements R_2 and C_2 make the frequency response of the feedback path flat over the desired frequency band (0.6 MHz), even with the increased bypass capacitor C_6. Due to larger C_6, the asymptotic losses A_T were further reduced. Note also that the physically shorter path for higher frequencies, through C_6 (but not through C_5 and C_6), reduces the loop phase lag.

Second, the Bode diagram was corrected in the step region by replacing the 20 Ω resistor between the bases of the output stage by the printed RLC two-pole. This corrector increased the stability margin. At the same time, due to its small resistance in the operational band, the loop gain was increased.

Third, the effects of arbitrary input and output termination on the high-frequency stability margins were eliminated by introducing the branches parallel to the amplifier input and output ports. Each branch represents the input of a loaded highpass filter with the limit frequency twice the highest frequency of the working band.

Fourth, an idle-winding input transformer was employed, the design of which is described in *problem 2* of Sec. 2.2.4.

Due to these changes, the operational bandwidth of the amplifiers was extended by 25%, and the linearity improved by 2 dB.

4. The feedback system of Fig. 2.81 having pure delay τ, i.e., nonminimal phase lag $|B_n| = \omega\tau$, in the feedback path approximates an ideal predictor (negative delay) over the bandwidth where the feedback is large and $K\beta = M/B_n = M/\tau\omega$. The typical responses are illustrated in Fig. 2.82. The larger the τ, the smaller is the bandwidth of available large feedback. Thus, a trade-off exists between the time of prediction and the accuracy of the prediction, determined respectively by the value and the bandwidth of the feedback.

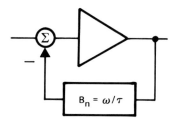

FIG. 2.81

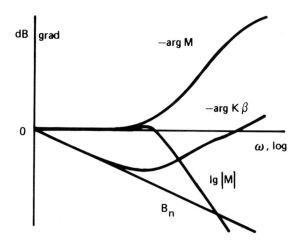

FIG. 2.82

2.5 NONMINIMUM PHASE SHIFT (NPS)

2.5.1 Causes for the Nonminimum Phase Shift

The time τ of signal propagation causes the frequency-proportional phase lag:

$$|B_n| = \omega\tau$$

At lower frequencies, such a loop phase is only noticeable if the length of the feedback loop is large or the speed of the signal in the medium is low (as with a thermal or acoustical signal). In a wideband system (a wideband feedback amplifiers or a phase-locked loop (PLL)), the propagation time of the electrical signal also constrains the available feedback. The delay time of signal propagation around the feedback loop having the average diameter D through a medium with the dielectric constant ξ is given by

$$\tau = \sqrt{\xi}\ \pi D/c$$

with c denoting the velocity of light. Then,

$$B_n = -2\pi f \sqrt{\xi}\ D/c$$

To keep $|B_{nc}|$ under 1 radian, as is desirable in a system with maximized feedback (recall Eq. (2.45)), the diameter D should be

$$D\ (\text{mm}) \leq 15/(\sqrt{\xi}\ f_c,\ \text{GHz})$$

Constantly improving technology lessens the physical dimensions of the feedback loops, thereby increasing the band of available large feedback.

The second main component of the NPS in the feedback loop is due to right-hand zeros of the transmission functions of the loop links.

The nonminimal phase shift produced by right-hand zeros s_i of a function $Q(s)$ can be calculated as follows. Introduce the function:

$$Q_n = \prod_i \frac{s + s_i}{s - s_i}$$

Since all the s_i are either real or constitute complex conjugate pairs [23], $|Q_n(j\omega)| = 1$.

Consider next the function $Q_m = Q/Q_n$. Evidently, $|Q_m(j\omega)| = |Q(j\omega)|$, i.e., $\ln|Q_m(j\omega)| = \ln|Q(j\omega)| = A(\omega)$, and $\theta_m = \ln Q_m$ is MP, so $B_m = \arg Q_n(j\omega)$ can be calculated from the response $A(\omega)$ with (2.29). The nonminimum phase shift $B_n = B - B_m$ is, therefore,

$$B_n = \arg Q_n(j\omega) = \sum_i 2\arg(j\omega - s_i) \tag{2.56}$$

In particular, each real right-hand-zero σ_i contributes the component of B_n:

$$B_{ni} = -2\arctan(\omega/\sigma_i) \tag{2.57}$$

As mentioned, $|B_n|$ is preferred to be under 1 rad in the band $\omega \leq \omega_c$. Then, from (2.56), it follows that in this band ω is at least two to three times smaller than the smallest of the zeros $|s_i|$. Hence, from (2.54), B_n is proportional to ω. For instance, for $\omega \leq 0.4\sigma_i$, i.e., for $|B_{ni}| \leq 0.8$ rad, (2.57) is closely approximated by

$$B_{ni} \cong -2\omega/\sigma_i$$

As noted in Sec. 2.2.1, such components of B_{nc} in the feedback loop are usually small. In the conventional cascade connection of passive ladder networks and ideal amplifiers, the transfer coefficient (or the transfer immitance) $Q(s)$ gets its zeros as the zeros of admittances of series branches, and the zeros of impedances of parallel branches, all being left-handed; thus $Q(s)$ becomes MP.

When a local loop is present with a stable local feedback path $\beta_l(s)$ and the forward path $Q_l(s)$ do not have right-hand poles, the transmission function of the link with the feedback $Q_l/(1 + Q_l\beta_l)$ has the same right-hand zeros as Q_l. Hence, such local feedback does not affect B_n.

2.5.2 Parallel Connection of Two Links

Using the consideration discussed in Sec. 2.14 and exploiting the Nyquist diagram for W_1/W_2, we can find the number of right-hand zeros of the transfer function $W_1 + W_2$ of a link composed of two parallel links W_1 and W_2. If, in particular, W_1 and W_2 are stable and MP, then the function $W_1 + W_2$ is MP as well if and only if the Nyquist diagram for W_1/W_2 does not encompass the point -1. Since with such functions W_1 and W_2, the ratio W_1/W_2 is also stable and MP, we can determine whether the Nyquist diagram encompasses the critical point by examining the Bode diagram for W_1/W_2.

In practice, it frequently happens that at $\omega = 0$, $W_1/W_2 < -1$; and at $\omega = \omega_b$ (where $|W_1/W_2| = 1$), $|\arg W_1/W_2| < \pi$; i.e., the average slope of the Bode diagram for W_1/W_2 is less than 24 dB/octave near ω_b. The Nyquist diagram for W_1/W_2 then accepts the shape shown in Fig. 2.83. The diagram makes half a revolution around the point -1 (the revolution would be complete if we were to draw the image of the whole $j\omega$-axis, from $-\infty$ to ∞). Hence, the sum $W_1 + W_2$ possesses a single right-hand zero. This zero is certainly real, and is readily found with conformal mapping, as was described in Sec. 2.1.2. Then, B_n is to be calculated with (2.55).

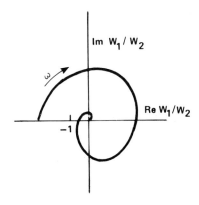

FIG. 2.83

As an example, consider the case shown in Fig. 2.84, where the locus of W_1/W_2 on the Nichols chart perpendicularly crosses the 0 dB level line. In

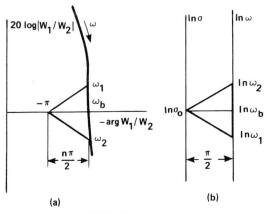

FIG. 2.84

the neighborhood of the crossover frequency ω_b, we approximate the function by linear dependence of $\ln W_1/W_2$ on $\ln s$:

$$\ln \frac{W_1}{W_2} = \ln\left[-\left(\frac{\omega_b}{s}\right)^n\right] = n \ln \omega_b - n \ln s - \pi$$

with n some real constant. Then, the regular triangle with the altitude $\pi/2$ of the $\ln s$-plane plots onto the regular triangle with altitude $n\pi/2$ in the plane $\ln W_1/W_2$. Due to logarithmic scales, $\omega_b = \sqrt{\omega_1 \omega_2}$, where ω_1 and ω_2 are associated with the summits of the triangle, so that $\ln \omega_2 - \ln \omega_1 = \sqrt{3}\,\pi/2$.

We see that the point σ_0, which plots onto the critical point $(-\pi, 0\,\mathrm{dB})$ in the $\ln W_1/W_2$ plane, is the real zero of W_1/W_2. As $\sigma_0 = \omega_b$, from (2.57),

$$B_n \cong -2 \arctan \frac{\omega}{\omega_b}$$

In the example shown in Fig. 2.85, the slope of the Bode diagram for W_1/W_2 is constant near ω_b, but $|\arg W_1/W_2|$ increases with frequency. In this case, $\sigma_0 = \omega_0 \leq \omega_b$.

Consider next the design problem of how to arrange the frequency responses for MP functions W_1 and W_2 in order to get $W_1 + W_2$ to be MP and robust, i.e., rather insensitive to variations of W_1 and W_2.

The sensitivities $1/(1 + W_1/W_2)$, $1/(1 + W_2/W_1)$ of the sum $W_1 + W_2$ to W_1 and W_2, respectively, become unlimited as the ratio W_1/W_2 approaches -1. To constrain them, we require from the hodograph of W_1/W_2 not to penetrate into a safety margin around the point -1. Analogously with the stability margins, we introduce the phase safety margin $y\pi$, and the amplitude safety margins x and x_1.

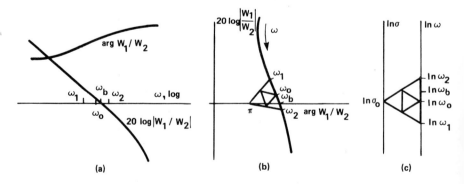

FIG. 2.85

The problem of connecting in parallel two links with adjoint transmission frequency bands, illustrated in Figs. 2.86 and 2.87, is of great practical significance. In Fig. 2.86, two amplifiers, the first with better performance at lower frequencies, and the second with better performance at higher frequencies, are connected in parallel by means of two forks of lowpass-highpass filters. The composed link is supposed to combine the best properties of the elementary links. The composed link is required to have a MP transmission function to allow itself to be looped by large feedback. The difference in the number of inverting stages in the two amplifiers must be even, otherwise the Nyquist diagram for W_2/W_1 encircles the critical point.

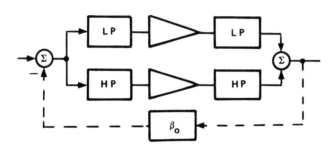

FIG. 2.86

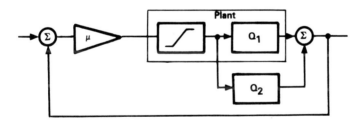

FIG. 2.87

Figure 2.87 presents the feedback system around the plant consisting of a nonlinear memoryless actuator and a linear lowpass link Q_1. The output of the actuator is accessible for measurement. The objective of the design is to attain maximum feedback within the working band. The high-frequency bypass Q_2 provides $|Q_1+Q_2|>|Q_1|$ at higher frequencies, thus permitting us to increase the feedback in the working band around the actuator without causing instability. However, the link Q_2 may be viewed as a local feedback path for the input amplifier μ and the actuator. This feedback reduces the gain and, therefore, the feedback around the link Q_1. Hence, the ratio $|W_1/W_2| = |\mu Q_1/\mu Q_2|$ is required to be as large as possible in the working band and, at the same time, as small as possible at frequencies beyond ω_c.

Figure 2.88 presents several examples of Bode diagrams for W_1 and W_2 and W_1/W_2. When the slope of the Bode diagram for W_1/W_2 is excessively large, the locus of W_1/W_2 encircles the critical point, thus indicating that the function $W_1 + W_2$ is not MP.

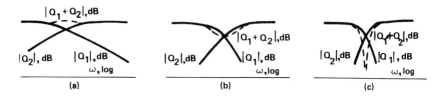

FIG. 2.88

Drawing the analogy from Bode analysis of the feedback system stability (recall Sec. 2.4.1, 2.4.2), the MP character of $W_1 + W_2$ will be preserved if the Bode diagram for W_1/W_2 near the intersection with the 0 dB level line is implemented as shown in Fig. 2.89(a). The difference in the slopes of the Bode diagrams for W_1 and W_2 at their intersection in Fig. 2.89(b) should be ~ 10 dB/octave. The diagrams should be parallel to each other over certain frequency intervals forming the upper and lower steps on the Bode diagram for W_1/W_2. The shorter the steps, the smaller is y and, therefore, the smaller is $(W_1 + W_2)/W_1$ at the intersection. If $y < 1/3$, then at the frequency of intersection $|W_1 + W_2| < |W_1|$, which yields a valley or a collapse on the Bode diagram for $W_1 + W_2$ as shown in Fig. 2.88(a,b,c).

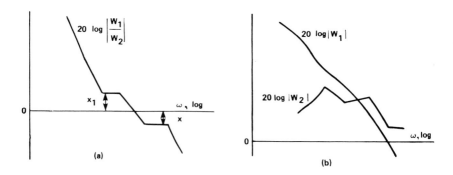

FIG. 2.89

The problem of achieving maximum selectivity of W_1/W_2 coincides with the problem (2.49) of attaining maximum feedback. Its solution is illustrated in Fig. 2.90(a,b).

If W_1 or W_2 (or both) is not MP, the problem arises of providing a minimum of $|B_n|$ for the composed path. This problem will be studied in Sec. 2.5.4.

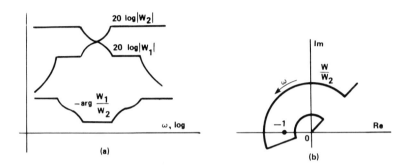

FIG. 2.90

2.5.3 Effect of Loading

The output voltage at the load Z_L of the two-port shown in Fig. 2.91 differs from the unloaded output voltage by a factor presented by the transfer function of a ladder network, comprising the output impedance of the two-port Z' and the load impedance Z_L. If Z_L and Z' are both positive-real functions, this factor is certainly MP, and the NPS B_n of the two-port transmission does not depend on Z_L. Similar analysis of a ladder network composed of the source impedance Z_1 and the input impedance of the two-port shows that in a passive circuit the phase shift B_n does not depend on Z_1.

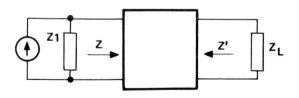

FIG. 2.91

It follows that when the impedances of the load, the source, and the input and output of the two-port are positive-real functions, then:

the result of calculation of B_n remains the same after replacing Z_L by any passive impedance (in particular, for the convenience of the analysis, by 0 or ∞);

the NPS of each of the transfer functions: $K = U_2/U_1$, $Z_{21} = U_2/I_1$, $K_I = I_2/I_1$, $Y_{21} = I_2/U_1$ is the same, and the MP character of either of the functions implies MP character of the others.

The freedom of taking each one of the functions and selecting an arbitrary passive load eases the analysis. For example, for a two-port composed of two parallel-connected four-poles having passive input and output impedances, excited by the source with a positive-real internal impedance and loaded at a passive two-pole, we can calculate B_n as the NPS of the sum of the two transfer admittances (of the output-to-input type) of the elementary four-poles, rather than calculating the two-port transmission function.

Let us apply this principle to the study of an amplifier stage with compound local feedback shown in Fig. 2.92. The input-output capacitance of the stage is denoted as C. The transfer admittance of the amplifier element y_{21} is assumed to be real and negative. It is reasonable to suppose that at zero frequency the amplification of the amplifier element is larger than the direct propagation through the feedback path, and therefore the total transfer admittance $Y_{21} = I_2/U_1 < 0$.

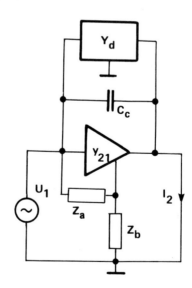

FIG. 2.92

In the simplified circuit, when $Y_d = 0$ and $Z_b = 0$, the transfer admittance $Y_{21} = Cs - |y_{21}|$ possesses a positive real zero:

$$\sigma_0 = y_{21}/C$$

and from (2.57) the NPS is [23]:

$$B_{n0}(\omega) = -2 \arctan \omega/\sigma_0 = \arctan \omega C/|y_{21}| \qquad (2.56)$$

At lower frequencies, $\omega < 0.4 C/|y_{21}|$:

$$B_{n0}(\omega) = -2C\omega/|y_{21}| \qquad (2.57)$$

Next prove the theorem: if $Z_b(s)$ is non-zero and positive-real, the transadmittance $\mathbf{Y}_{21}(s)$ is non-zero and MP, and $\mathbf{Y}_d(0) > 0$, then the NPS $B_n > B_{n0}$.

As the local feedback under the condition of shorted input and output (recall Eq. (1.32)) is

$$F_l(0) = \frac{y_{21} Z_a Z_b}{Z_a + Z_b} + 1$$

then

$$Y_{21} = Y_d + Cs + \frac{\mathbf{y}_{21}}{F(0)}$$

or

$$Y_{21} = Y_d + Cs - \frac{1}{1/\mathbf{y}_{21} + Z_a Z_b/(Z_a + Z_b)} \qquad (2.58)$$

On the positive-real semiaxis, $\sigma = \operatorname{Re} s > 0$, all the functions entering (2.58) are real, Z_b as well as the two first components are positive, and the third component is negative. Inspection of (2.58) reveals that Y_{21} increases monotonically with increase of Y_d or Z_b. Therefore with nonzero Y_d or Z_b, the value of $Y_{21}(\sigma_0) > 0$. Hence, $Y(\sigma)$ changes its sign on the interval $(0, \sigma_0)$. Because there cannot be poles on this interval due to the finite character of the components, a zero should exist $\sigma_1 \in (0, \sigma_0)$. Therefore,

$$B_n = -2 \arctan (\omega/\sigma_1) > B_{n0} \qquad (2.59)$$

We see that direct propagation through the local feedback circuits may increase $|B_n|$ and, respectively, reduce the available common feedback. Fortunately, such a $|B_n|$ is typically small, permitting the local feedback to be used for sensitivity reduction of the looped link, or frequency response correction of the main-loop transmission function.

A proper choice of the type of local feedback can appreciably reduce $|B_n|$. For instance, if the load and source impedances for the stage are low, the impedance of the two-pole of the parallel local feedback path must be also low, which causes larger direct signal propagation and NPS than series local feedback with comparable value of feedback. Conversely, with high-impedance source and load, parallel feedback is preferred.

Estimation of B_n is rather complicated in a cascade connection of active two-ports, where the input and output impedances are not necessarily positive-real and B_n depends on the loading conditions. This situation arises when cascading feedback amplifiers as active links of the common loop. Fortunately, the two following conditions are often satisfied, allowing us to simplify the analysis by replacing the load impedance Z_L by a fictitious load with the impedance Z (conveniently, 0 or ∞):

- a. The link should be stable while loaded at Z_L, and remain stable after changing the load to Z, so that this replacement does not change or create right-hand poles of Q.
- b. The load impedance should not have right-hand zeros if the voltage U_2 is the measure of the output signal, and it should not have right-hand poles if the output signal is observed as the current I_2. The need for meeting this condition is shown below.

If a voltage is understood to be the output, the load impedance contributes its zeros to the considered transfer function (K or Z_{21}). Therefore, as the right-hand zeros of the function may not be affected by replacement of Z_L by Z, these two impedances must possess the same right-hand zeros. A case of practical interest is when neither Z_L nor Z have such zeros, i.e., when the two-poles with such impedances are stable when shorted. Similarly, if a current is understood to be under the output, the two-poles must be stable when open (disconnected).

In the ordinary case when the local feedback is either parallel or series (not compound) at the links' inputs and outputs, the analysis can be effectively simplified using the two following recommendations:

- c. Dimensionality of the signal at the links junctions should be chosen such that the effect of the local feedback can be neglected when calculating the transfer function of the following link.
- d. The impedance of the equivalent load should be finite, constant, real, and either very large or very small so as to simplify the analysis of the direct transmission through the followed link.

For example, if the local feedback in the followed link is of voltage type (parallel), and in the following link of current type (series), then the signal at the junction should be taken as voltage, and a very small resistance should play the role of the equivalent load.

This matter is illustrated with the three-stage feedback amplifier shown in Fig. 2.93, which is supposed to be stable without the main loop. In accordance with recommendation (c), the dimensionality of the signals at the links' junctions is chosen as I_1, U_2, I_3, I_4 (or U_4). As equivalent loads, very large resistances, R_2 and R_4, and very small resistance, R_3, have been chosen. As the result, conditions (a) and (b) are satisfied for all the links, and B_n for each link can be calculated with equivalent loads, thus greatly simplifying the analysis.

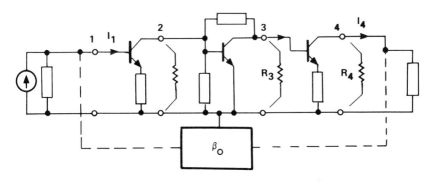

FIG. 2.93

2.5.4 Noncascade Connections of Active Two-Ports

Under the noncascade connection, we understand parallel, series, parallel-series, and series-parallel configurations, and, with the former as particular cases, the configuration of Fig. 2.94, where the two-ports are connected by means of input and output six-poles. The signal proceeds from the input to the output along two paths, and their transfer coefficients (immittances) μ_1 and μ_2 are summed.

Let μ_1 include a larger number of amplification stages (assumed to be integrators at higher frequencies) than μ_2. Accordingly, in the operating frequency band $|\mu_1| > |\mu_2|$, but at frequencies $\omega > \omega_{Tav}$, conversely, $|\mu_2| > |\mu_1|$. The total transfer coefficient $\mu = \mu_1 + \mu_2$ shares the best properties of both paths, since $\mu \cong \mu_1$ in the operating band and $\mu \cong \mu_2$ at asymptotic frequencies.

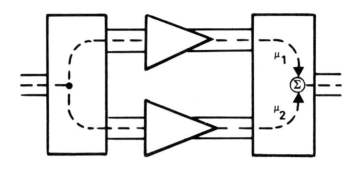

FIG. 2.94

The quantities μ_1 and μ_2 include as factors the transfer coefficients (immittances) of the input and output six-poles, i.e., of the signal splitter and combiner.

The splitter is expected to transmit most of the energy of the input signal at frequencies $\omega \leq 1$ to the input of the first amplifier, and in the frequency range near ω_c to the input of the second amplifier. For this reason, the splitter should represent a fork of lowpass and highpass filters with the separation band $[1, \omega_c]$ and the attenuation in the stop-bands of 10 to 20 dB. Analogous requirements are imposed on the combiner.

The filter's complexity depends on the loading conditions. In particular, these may be simple RC-links (when, for example, a high-input-impedance operational amplifier is used as μ_1 and a wideband amplifier as μ_2), or much more complicated RLC circuits.

The filter's selectivity is restricted by the requirement for the NPS of the total channel to be kept as small as possible, i.e., equal the smaller of the NPS of μ_1 and μ_2 (which is apparently the latter).

As the number of integrators in the path μ_1 is larger than in the path μ_2,

$$\lim_{\omega \to \infty} \frac{\mu_1}{\mu_2} = 0$$

Hence, the integral of the real part of the function (2.10) can be applied to the function:

$$\theta = \ln \mu - \ln \mu_2 = \ln (\mu / \mu_2) = \ln (1 + \mu_1 / \mu_2) \tag{2.61}$$

giving

$$\int_0^\infty (\ln|\mu| - \ln|\mu_2|)\, d\omega = -\frac{\pi}{2} B_\infty \qquad (2.62)$$

Expanding $\ln(1 + \mu_1/\mu_2)$ at high frequencies in a series of positive powers of μ_1/μ_2 and confining ourselves to the first term only, we obtain

$$B_\infty = \lim_{\omega \to \infty} j\omega\, \mu_1/\mu_2 \qquad (2.63)$$

which is connected with the position of high-frequency asymptotes of μ_1 and μ_2. These asymptotes are given by the expressions $(\omega_{T1}/j\omega)^{n_1}$ and $(\omega_{T2}/j\omega)^{n_2}$, respectively, with ω_{T1} and ω_{T2} designating the frequencies at which $|\mu_1|$ and $|\mu_2|$, respectively, reduce to 1. Then

$$B_\infty = \begin{cases} 0: & n_1 - n_2 = 2 \\ \omega_{T1}/\omega_{T2}: & n_1 - n_2 = 1 \end{cases} \qquad (2.64)$$

If, in particular, the gain of the amplifier stages of both amplifiers become 0 dB at the same frequency ω_T, and at frequencies near ω_c the product of transfer coefficients of the splitter and the adder related to the first channel is the constant k_{sa1c}, then

$$\omega_{T2} = \omega_T$$
$$\omega_{T1} = \omega_T (\mu_{sa1c})^{1/n_1}$$
$$B_\infty = -\omega_T k_{sa1c} \qquad (2.65)$$

If $|\mu_{sa1c}|$ is fairly small, the right-hand side of (2.62) can be neglected, as well as in the case of $n_1 - n_2 > 1$.

Relationship (2.62) states that the differences in gain between the combined path and the path μ_2 is limited in area. Thus, if $|\mu| \cong |\mu_1|$ grows excessively large in the working band $\omega = 1$, it will wade at other frequencies below the desirable loop gain response, which provides maximum feedback. This fact constrains $\ln|\mu_2|$ from below throughout the working band, and thereby limits from below the n_2 and the asymptotic slope of the response of $\mu = \mu_2$ at higher frequencies, i.e., restricts the available feedback.

The maximum achievable $a = \ln|\mu| - \ln|\mu_2|$ over the working band can be roughly evaluated as follows. With only the amplifier μ_2 on, the Bode diagram for the return ratio must be similar to curve 1 in Fig. 2.95(a). For the return ratio with the two amplifiers on, implemented as the optimum Bode cut-off shown by curve 2, piecewise approximation of the difference between these responses is shown in Fig. 2.95(b). For this approximation, (2.62) changes to

$$a + \int_1^{\omega_q} \left(a + \frac{a-q}{\ln \omega_q} \ln \omega \right) d - \int_{\omega_c}^{\omega_q} \frac{q \ln \omega_c + q \ln \omega}{\ln \omega_c - \ln \omega_q} d\omega = -\frac{\pi}{2} B_\infty \qquad (2.66)$$

where q and ω_q are as shown in Fig. 2.91.

The integrands are linear functions of $\ln \omega$. Since

$$\int \ln \omega \, d\omega = \omega \ln \omega - \omega + \text{const},$$

by integrating and collecting the terms, we obtain

$$a \frac{\omega_q - 1}{\ln \omega_q} - q \left[\frac{\omega_c - \omega_q}{\ln \omega_c - \ln \omega_q} + \frac{\omega_q - 1}{\ln \omega_q} - 2\omega_q \right] = -\frac{\pi}{2} B_\infty \qquad (2.67)$$

This equation is linear with respect to the parameters a and q, so these are easily determined, provided that the remaining parameters are given.

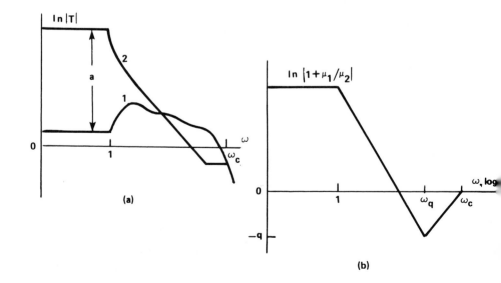

FIG. 2.95

Several examples of noncascade configurations of transistor circuits are shown in Fig. 2.96. The simplest circuit (a) is advisable for use when, for example, Q_1 is a high-power transistor and Q_2 is a lower-power transistor with higher f_T, and their combination serves as the final amplification stage. (This system must employ a nonlinear dynamic compensator, described in Chapter 5, since otherwise overloading of the low-power transistor at high frequencies may create the conditions for self-oscillation.)

In the circuit of Fig. 2.96(b), $n_1 = 3$, and $n_2 = 1$. Therefore, introducing the parallel path through the transistor Q_2 increases the available feedback, even if this transistor has no better frequency response than the others. The shortcomings of this circuit are its necessity in complicated networks L_1 and L_2 (which in the realization of an appropriate Nyquist diagram contribute to the ratio μ_1/μ_2), and rather stringent requirements for the accuracy of the frequency responses of μ_1 and μ_2.

In the circuit of Fig. 2.96(c), the path μ_1 includes two common-emitter (CE) transistor stages with transistors Q_1 and Q_2, while the path μ_2 includes transistor Q_1 as a common-collector (CC) stage and transistor Q_3 as a common-base (CB) stage, both paths being noninverting. In this circuit,

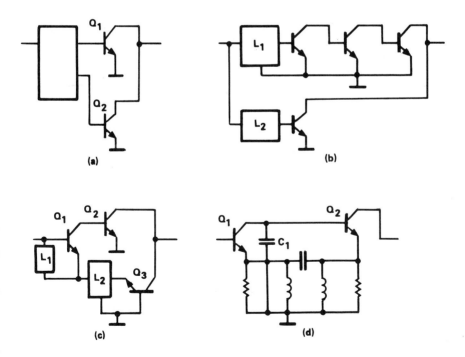

FIG. 2.96

$n_1 = 2$ and $n_2 = 1$, since the current gain of CC-CB circuit configuration is known to be equal to the gain of a single CE stage. The networks L_1 and L_2 play the role of a fork of filters.

The circuit of Fig. 2.96(d) exhibits similar properties. It reduces to a CE-CE configuration for $\omega \leq 1$ and to CC-CB configuration at frequencies in the neighborhood of ω_c, so that $n_1 = 2$ and $n_2 = 1$. The typical experimental response for the current gain A and the related phase shift B are shown in Fig. 2.97.

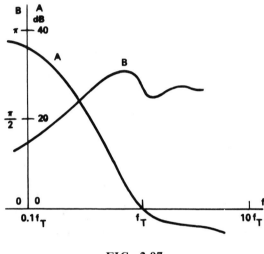

FIG. 2.97

Chapter 3
System with Single Nonlinear Link

3.1 ABSOLUTE STABILITY PROBLEM

A. Lurie's problem of absolute stability [94, 95 *see also* 6] is to find the conditions for asymptotic global stability (AGS) of a linear link $-T_0(s)$ looped by a memoryless (i.e., nondynamic nonlinear link $v(e)$, $0 < dv/de < r$.

For the sake of simplicity, we normalize the transmission coefficients by making $r = 1$. The normalized system is said to be absolutely stable (AS) if it is AGS with any characteristic $v(e)$ constrained by

$$0 < \frac{v(e)}{e} < 1 \qquad (3.1)$$

as illustrated in Fig. 3.1.

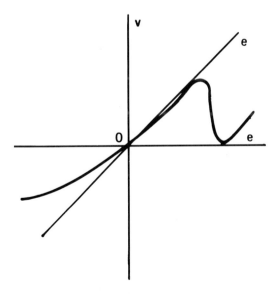

FIG. 3.1

The criterion for the AS proposed by A. Lurie [94] employs the second Liapunov method of stability analysis. It supplies sufficient stability conditions that in practice are very close to the unknown necessary conditions, especially for systems with typical lowpass responses of the linear link.

The mathematics of A. Lurie [94, 95] is overly complicated for everyday engineering. Instead of using it, engineers took for granted that a system with typical linear links should be AGS if the system is stable after replacing the nonlinear link by any memoryless linear link k with $0 < k < 1$. The conjecture of M. Aizerman that this statement is generally correct was repudiated by V. Pliss who constructed a counterexample with a third-order linear link having a bandpass characteristic [154], and thus showed that the problem of absolute stability is consistent.

The frequency-domain absolute stability criterion due to V.M. Popov [6, 75, 137, 143, 172] was proved by R.E. Kalman [82] to be equivalent in strength to A. Lurie's method. The Popov criterion is, fortunately, much easier both to apply and to teach. It will be discussed in the following sections.

The AS criterion is generalized (although it is rarely required in practice) on systems with the characteristic $v(e)$ constrained by two rays. By normalizing such a characteristic to the form shown in Fig. 3.2 and adding to the system two mutually cancelling branches $-r$, as indicated in Fig. 3.3, we

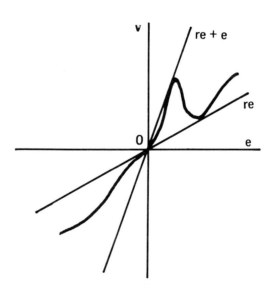

FIG. 3.2

System with a Single Nonlinear Link

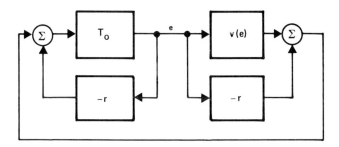

FIG. 3.3

obtain a new system whose return ratio is $T_0/(1+rT_0)$ and whose nonlinear link characteristic $v(e) - re$ belongs to the sector shown in Fig. 3.1. Therefore, we may limit ourselves without losing the generality to the study of systems with the characteristic $v(e)$ satisfying inequality (3.1). The methods for the system synthesis, which will be developed in Chapter 8, will use only such nonlinear links.

3.2 POPOV CRITERION

3.2.1 Nonlinear Physical Two-Poles

The Popov absolute stability criterion may be understood through the analogy between the feedback system and the connection of two-poles (refer to Sec. 1.1).

Consider the nonlinear RL two-pole of Fig. 3.4(a) with the nonlinearity of the resistor and the inductor obeying the same nonlinear law, so that the voltage applied to the terminals of the two-pole is

$$u = \Phi(i) + \frac{dq\Phi(i)}{dt} \tag{3.2}$$

where i is the current, Φ is the voltage drop across the resistor, q is a certain positive constant, and $q\Phi$ is the magnetic flux produced by the inductor. The circuit diagram of the same two-pole is redrawn in Fig. 3.4(b) with a linear resistor of $1\,\Omega$ detached from the rest of the circuit, so that

$$u = i + \Phi(i) - i + \frac{dq\Phi}{dt}$$

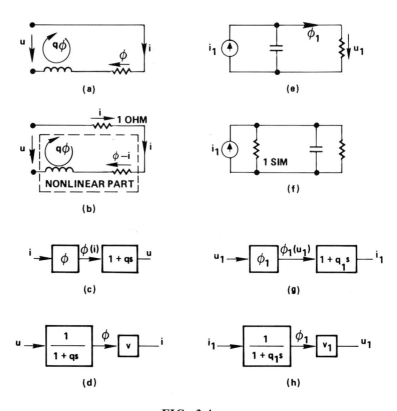

FIG. 3.4

Next, assume $\Phi/i > 1$ as is shown in Fig. 3.5(a). Then, the voltage drop $\Phi(i) - i$ and the power $[\Phi(i) - i]i$ dissipated on the nonlinear resistor are positive for all i. Hence, this resistor is inherently passive. Its voltage-current characteristic, illustrated in Fig. 3.5(b), can be physically implemented by interconnection of only passive elements such as linear resistors, diodes, tunnel diodes, *et cetera* for any prescribed interval of i.

The inductor producing the flux $q\Phi$ can also be realized as a passive physical system, for example, by the electromechanical arrangement drawn in Fig. 3.6(a).

Cascade connection of a memoryless nonlinear link Φ and the linear link $1 + qs$ depicted in Fig. 3.4(c) is described by the same formula (3.2). Figure 3.4(d) shows the diagram for the inverse direction of signal propagation, with the memoryless link operator v inverse with respect to Φ. Note that the operator v satisfies the condition (3.1), as can be seen in Fig. 3.5(a).

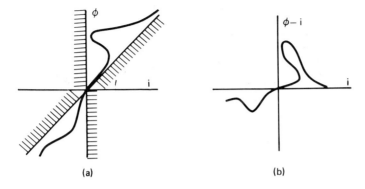

FIG. 3.5

The circuits dual with respect to those of Fig. 3.4(a,b), i.e., described by the same equations after interchanging the voltage and the current, are the RC circuits shown in Fig. 3.4(e,f). The nonlinear conductance can be realized using tunnel diodes and other passive elements without energy sources, and the nonlinear capacitor, by the electromechanical system illustrated in Fig. 3.6(b).

The interconnection of the above-described two-poles with some other passive linear and physical passive nonlinear two-poles creates a physical passive system, which is, therefore, stable.* The attendant closed-loop flow chart describes a particular equivalent feedback system, which must be stable as well. This analogy generates a variety of stability criteria for feedback systems.

3.2.2 Popov Criterion

Let us couple the two-pole of Fig. 3.4(a,b) to a two-pole having impedance $Z(s) - 1$, where $Z(s)$ is positive-real. The resulting circuit is realizable using physical passive elements only, and is therefore AGS. Using Fig. 3.4(d), the circuit flow chart is obtained from Fig. 3.7. This feedback system is certainly AGS as well. The loop transmission function of the linear links of the loop is

$$T_0(s) = [Z(s) - 1]/(1 + qs) \tag{3.3}$$

---

*A system constructed of components satisfying some of the diverse definitions of "passivity" [175] is not necessarily AGS. This is why we use the concept of physical passive elements, i.e., the components with which a *perpetuum mobile* cannot be built.

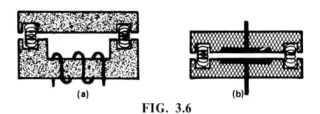

FIG. 3.6

from which, we have

$$Z(s) = (1 + qs) T_0(s) + 1 \tag{3.4}$$

Therefore, a feedback system consisting of a nonlinear link v and the linear link T_0 is AGS if $T_0(s)$ can be presented in the form (3.3) with positive-real $Z(s)$, i.e., if such positive real q exists that makes expression (3.4) positive-real, i.e.,

$$\text{Re}\,(1 + jq\omega)\,T_0\,(j\omega) > -1. \tag{3.5}$$

This statement constitutes the Popov absolute stability criterion.

Adding qs to $Z(s)$ does not affect the positive-real condition [23]. Then, the second, slightly different form of the criterion states that the system is AGS if a real positive q exists with which

$$(1 + qs) F_0(s) \text{ is positive-real} \tag{3.6}$$

i.e., at all frequencies,

$$|\arg[(1 + jq\omega) F_0\,(j\omega)]| < \pi/2 \tag{3.7}$$

The AGS, therefore, may be verified either by varying q so as to satisfy, if possible, the inequality:

$$\arg(1 + jq\omega) + \arg F_0 < \pi/2 \tag{3.8}$$

at all frequencies, or by plotting the Bode diagram for $(1 + jq\omega) F_0\,(j\omega)$ and, considering its shape, making the judgement of whether (3.7) is satisfied by using the phase-gain functional.

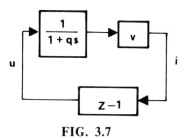

FIG. 3.7

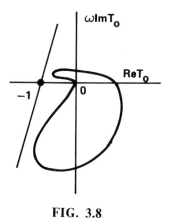

FIG. 3.8

Figure 3.8 illustrates an application of the third form of the Popov criterion, which uses the plane of the *modified function* $\operatorname{Re} T_0(j\omega) + j\omega \operatorname{Im} T_0(j\omega)$. Equation (3.5) can be rewritten as

$$-1 - q\omega \operatorname{Im} T_0(j\omega) < \operatorname{Re} T_0(j\omega) \tag{3.9}$$

If made equal to $\operatorname{Re} T_0(j\omega)$, the left side of the inequality represents a Popov line. It passes through the point -1 with the slope coefficient of $-1/q$ on the plane of the modified function. Therefore, inequality (3.9) is satisfied, and the system is AS if a Popov line can be drawn to the left from the locus of the modified function. For example, the system for which the locus of the modified function is plotted in Fig. 3.8 is AS.

The system characterized by the locus shown in Fig. 3.9 can neither be said to be stable nor unstable. This plot only guarantees the absence of certain types of oscillation. No periodic oscillation, for example, may be present if a Popov line can be drawn to the left of all the points on the locus of the modified function that relate to the frequencies of the Fourier components of the supposed oscillation [51]. Thus, periodic oscillation with fundamental ω_f cannot exist in the system represented in Fig. 3.9. Applying this method with a large value of q to a passband feedback system with a stable type of Nyquist diagram, it is usually possible to exclude all types of periodic oscillation as being impossible, except those with a fundamental belonging to the frequency range of low-frequency roll-up, where $\arg T_0(j\omega) > \pi/2$.

The Popov criterion is generalized in several diverse aspects. It is proved, in [28], for example, that the criterion is valid with normalized function of any RL passive impedance instead of the Popov factor $(1 + qs)$.

The matrix Popov criterion [7, 138] is applicable to the analysis of certain systems containing several nonlinear links of the same kind. However, the matrix form lacks the transparency of the scalar form, especially for systems having optimized frequency responses. This is why synthesis procedures utilizing the AS criterion, to be proposed in the following chapters, are limited to those systems with several memoryless nonlinear links, which are reducible, by some equivalent transformation, to systems with a single nonlinear link.

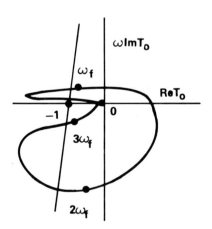

FIG. 3.9

3.2.3 Applications

A lowpass system with Bode optimal cut-off is AS. This can be established with the second form of the Popov criterion when employing the RL function of the type $(1+q_1s)/(1+q_2s)$ instead of the Popov factor. With $1/q_1 \ll 1$ and $1/q_2 = 20\,\omega_c$, the modified inequality of (3.7), expressed as $|\arg F_0 + \arg(1+j\omega q_1)/(1+j\omega q_2)| < \pi/2$, is satisfied, as can be seen in Figs. 3.10 and 3.11. We can also see that with $q_2 = 0$, i.e., using the regular Popov factor, the inequality cannot be satisfied because q_1 is required to be large to meet (3.7) at $\omega = 1$, and at the same time it is required to be small to satisfy (3.7) at $\omega = 4\,\omega_c$, where $\arg F_0 > 0$.

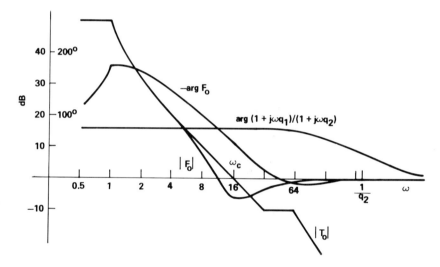

FIG. 3.10

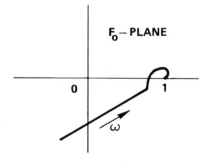

FIG. 3.11

Next, consider a bandpass system with the feedback maximized over a large relative frequency band, having the phase stability margin $y\pi$ in the region of high-frequency cut-off, and $y_l\pi$ in the region of lower-frequency roll-up, as shown in Figs. 3.12 and 3.13. We can see that the combination of $y=0$ and $y_l=0.5$ satisfies the condition (3.7) at its limit, i.e., $\arg(1+jq\omega)T_0$ varies from $-\pi/2$ to $\pi/2$, if $1/q$ is chosen to be close to the mean square frequency of the passband. Deviations to either side of this value worsen the situation: reducing q reduces $\arg(1+jq\omega)T_0$ at the high-frequency cut-off such that it falls below $-\pi/2$; making q larger increases this argument at the lower frequencies of the roll-up.

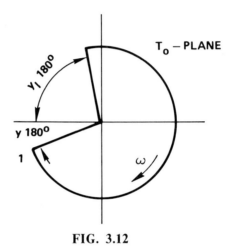

FIG. 3.12

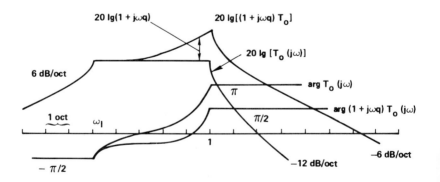

FIG. 3.13

Therefore, the system stability can be proved with the Popov criterion when only $y_l \geq 1/2$. However, the value of $y_l \cong 1/6$ is doing well in practical design. Making y_l excessively large to satisfy the criterion imposes little impairment of the available feedback, as can be understood by using the phase integral (2.18), but requires an extension of the loop gain to still lower frequencies.

In narrowband passband systems, the Popov sufficient stability criterion requires that the roll-up and cut-off phase stability margins, $y_l\pi$ and $y\pi$, exceed $\pi/2$. Satisfying this requirement would dramatically reduce the maximum available feedback, as compared with using a conventional stability margin of $\pi/6$ in the bandpass optimal cut-off.

The smaller the relative bandwidth, the smaller is the feedback limited by the Popov criterion because of deficient filtering properties of the Popov factor $(1+jq\omega)$. The RL functions that can be employed instead of the Popov factor have even weaker frequency selectivity and do not improve the criterion in this sense.

In practice, the phase stability margins of $\pi/6$ both at higher and lower frequencies are sufficient for the AGS of a system with saturation link, regardless of the relative bandwidth.

In conclusion, the available feedback depends on the location of the working band on the frequency scale, contrary to bandpass frequency transformation of the optimal baseband system, but not in quantitative agreement with the Popov criterion. Experimental and theoretical analysis of such a system is performed in Sec. 3.4.

3.2.4 Examples and Exercises

1. The Popov criterion with $q=0$ reduces to the so-called circle criterion [159, 170]. Consider how the fulfillment of this criterion will affect the available feedback in lowpass and bandpass systems.

2. Use the Popov criterion to prove that in a bandpass system with maximized feedback, having $y_l = y > 0$, no periodic oscillation can exist at frequencies higher than the mean square frequency of the passband.

3. Construct the Bode plot for T_0 to maximize the feedback in a passband system with $y_l = y = 0.3$, and having the relative bandwidth of 0.5 octaves, if the maximum available feedback in the lowpass prototype is 40 dB, for two versions: (a) on the basis of Aizerman's conjecture, and (b) satisfying the Popov criterion. Compare the obtained values.

4. The feedback loop in Fig. 3.14 consists of two nonlinear memoryless links v and v_1, each satisfying the condition (3.1), separated by two integrating links $1/(1+qs)$, $1/(1+q_1s)$. Prove the system AGS by using the mathematical analogy between this feedback system and the parallel connection of two two-poles shown in Fig. 3.4(a,e).

5. In the bandpass feedback system, the Popov criterion imposes a stronger restriction on the $T_0(j\omega)$ response than Aizerman's conjecture, apparently because of the interference between the low-frequency and high-frequency components in the nonlinear link (to be studied in Sec. 3.4). Then, it might be expected that such a system could be proved to be AS if the nonlinear link is replaced by two similar links connected in parallel through lowpass and highpass filters, thus reducing the intermodulation of the low-frequency and high-frequency components, as shown in Fig. 3.15. This system can be equivalently transformed into that of Fig. 3.16 with lowpass-highpass filter forks at the input and output of the nonlinear links.

To find the stability conditions for these systems, we consider Fig. 3.15 to be the flow chart for the circuit of Fig. 3.17 representing a series connection of three two-poles: of those shown in Figs. 3.4(a,b), and a linear one with the impedance $Z-1$. The circuit is certainly stable if $Z(s)$ is positive-real. Therefore, the attained feedback system of Fig. 3.15, with the linear links return ratio:

$$T_0 = (Z-1)\left(1 + qs + \frac{1}{1+q_1s}\right)$$

is AS if positive-real numbers q and q_1 exist such that the function:

$$Z(s) = T_0 / \left(1 + qs + \frac{1}{1+q_1s}\right) + 1$$

is positive-real.

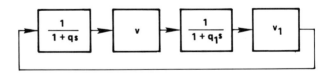

FIG. 3.14

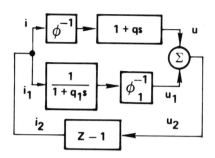

FIG. 3.15

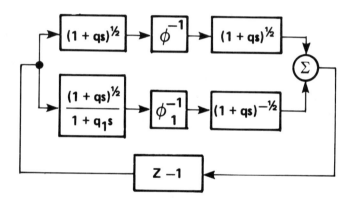

FIG. 3.16

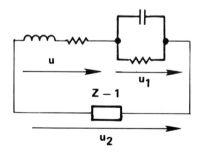

FIG. 3.17

3.3 Periodic and Nonperiodic Self-Oscillation

As a matter of experience, nonperiodic oscillation does not arise in practical systems of moderate complexity near the border of global stability. If we continuously vary the system parameters, departing from the inside of the region of global stability, then the first type of self-oscillation that can be excited is periodic.

Unable to prove this assertion, we will take it as a working hypothesis. We can only illustrate with an example that the parameters of a system exhibiting chaotic behavior are not typical for practical systems.

Figure 3.18(a) shows an implementation of Chua's chaotic attractor [131]. Presenting the nonlinear resistor as the series connection of a linear resistor and a modified nonlinear resistor yields the circuit of Fig. 3.18(b). As shown in Fig. 3.18(c), the equivalent feedback system with saturation nonlinear link possesses a return ratio that is infinitely increasing with frequency, which is the opposite of what happens in real control systems.

Employing the transformation depicted in Fig. 3.3 with $r = 1$ to the system of Fig. 3.18(c) yields an equivalent highpass system with finite loop gain and a dead-zone link (without saturation) in the loop. This system as well is not practical.

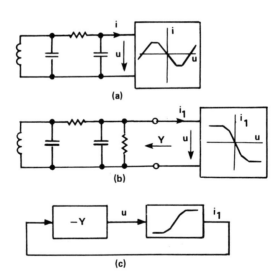

FIG. 3.18

3.4 MULTIFREQUENCY OSCILLATION IN A BANDPASS SYSTEM WITH SATURATION

3.4.1 Goals for Analysis

As mentioned, the generalized Popov criterion gives a sufficient condition of absolute stability that is, in fact, necessary for lowpass systems, but far from being necessary for bandpass systems. It is interesting, then, to study the stability conditions for bandpass systems, and particularly for those with maximized feedback having the saturation type of nonlinear link. It is instructive, further, to contemplate the interrelationship between the phase stability margin y_l at lower frequencies and the relative bandwidth of the working band, since the Popov criterion stability requirement for $y_l \geq 1/2$, being irrelevant to the relative bandwidth, is far more excessive than necessary.

It follows from experiments that the stability of bandpass systems with a saturation link in the loop is only conditional if the phase stability margin is too short, both at higher and lower frequencies. Such systems burst into stable periodic or quasiperiodic oscillation after temporary input of a periodic signal of a certain frequency range and sufficiently large amplitude. The oscillation stops abruptly when we gradually introduce changes in the linear part of the system so as to monotonically reduce the phase stability margins at higher or lower frequencies.

Due to unavoidable ripples in the practical frequency responses for T_0 and arg T_0, the frequency response for $y\pi = \pi - \arg T_0$ is not monotonic in the regions of roll-up and cut-off, and the fundamental of the oscillation is commonly located close to one of the local maxima of arg T_0, somewhere below the working band (recall Sec. 3.2.2). In the following sections, we will examine two kinds of the oscillation:

with ω_f located below, but close enough to, the working band $[\omega_l, 1]$, so that $|T_0(j\omega_f)| \gg 1$;

with ω_f located at much lower frequency, where $|T_0| \cong 1$.

3.4.2 Oscillation with Fundamental at which the Loop Gain is Large

If the self-oscillation is periodic, the transmission coefficient around the loop is 1 for each Fourier component of the oscillation, and particularly for the fundamental ω_f. In the following analysis, we assume $|T_0(j\omega)| \gg 1$; then, in order to satisfy the foregoing statement, the nonlinear link must be saturated during the greater portion of the period.

In accordance with observations of such oscillation in experimental devices [111], we suppose that the shape of $v(t)$ is close to trapezoidal, as shown in Fig. 3.19, with the clipping angle χ approaching $\pi/2$. The waveform is described by the equation

$$v(t) = \begin{cases} e(t): & \omega_f t \in [k\pi/2 - \chi, k\pi/2 + \chi] \\ v_s, & \text{otherwise} \end{cases}$$

where v_s is the amplitude, and k is any odd integer.

The waveform is approximated by the trapezium, such that at $\omega_f t = (k-1)\pi$,

$$e(t) = v(t) = 0 \tag{3.10}$$

$$\frac{de(t)}{d\omega_f t} = \frac{dv(t)}{d\omega_f t} = \frac{v_s}{\pi/2 - \chi} \tag{3.11}$$

Fourier series for $v(t)$ and $e(t)$ follow as

$$v(t) = \frac{2v_s}{(\pi/2 - \chi)} \sum_{k=1,3,5,\ldots}^{\infty} k^{-2} \sin k(\pi - 2\chi) \sin k\omega_f t$$

$$e(t) = -\frac{2v_s}{(\pi/2 - \chi)} \sum_{k=1,3,5,\ldots}^{\infty} k^{-2} \sin k(\pi - 2\chi) \operatorname{Re} T_0(k\omega_f) \sin k\omega_f t$$

$$+ \operatorname{Im} T_0(k\omega_f) \cos k\omega_f t \tag{3.12}$$

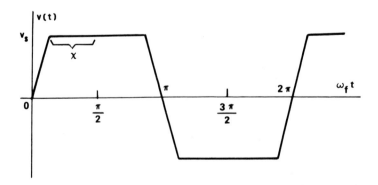

FIG. 3.19

By substituting the expression for $e(t)$ into (3.10) and (3.11), and omitting some simple algebra, the two necessary conditions for the self-oscillation are

$$\sum_{k=1,3,5,\ldots}^{\infty} k^{-2} \sin k(\pi - 2\chi) \operatorname{Im} T_0(k\omega_f) = 0 \qquad (3.13)$$

and

$$\sum_{k=1,3,5,\ldots}^{\infty} k^{-1} \sin k(\pi - 2\chi) \operatorname{Re} T_0(k\omega_f) = -\pi/2 \qquad (3.14)$$

Examine (3.13) first. Since χ approaches $\pi/2$, the weighting function $k^{-2} \sin k(\pi - 2\chi)$ rapidly decreases with k, and can be truncated after $k = \pi/(\pi - 2\chi)$. Up to this value of k, the weighting function is positive. Therefore, (3.13) may be satisfied if and only if $\operatorname{Im} T_0$ changes sign at some frequency below $\omega_f \pi/(\pi - 2\chi)$. This could happen only in a bandpass system (in agreement with Popov criterion).

Assume, therefore, that the system is bandpass with the normalized passband from ω_l to 1. In such a system, the sign of $\arg T_0$ changes at the center frequency of the passband $\sqrt{\omega_l}$. Thus, $\sqrt{\omega_l} < \omega_f \pi/(\pi - 2\chi)$.

Since the fundamental ω_f must belong to the frequency range where the phase stability margin is small, i.e., it must be less than ω_l, it follows that the inequality $\omega_l < \omega_l \pi/(\pi - 2\chi)$, or

$$\omega_l > (1 - 2\chi/\pi) = v_s/2\pi T_0(j\omega_f), \qquad (3.15)$$

is the necessary condition for the self-oscillation.

Consider next the relationships between the stability margins at higher and lower frequencies when the system faces self-oscillation. With y growing larger, the loop phase lag reduces; therefore, the frequency at which $\operatorname{Im} T_0$ changes sign increases. Hence, the components of the expression on the left side of (3.13) become larger as well. Then, in order to continue satisfying the equation, $\operatorname{Im} T_0$ must be reduced at lower frequencies, i.e., y_l should be lowered. Thus, an increase in y_l can be traded off for an increase in y, and vice versa.

A similar conclusion follows from the analysis of inequality (3.14). If y_l and y are both large, the left-hand side of (3.14) is positive. Reducing y_l or y makes the sum on the left side of (3.14) smaller due to frequency components located at the low-frequency roll-up and high-frequency cut-off.

The theory agrees well with the experimental findings (for more details, see references [111, 113]. Although the oscillation conditions found are only sufficient, they are probably close to the necessary ones. Quantitatively, the calculated and measured values of the phase stability margins at the edge of self-oscillation are much smaller than those required by the Popov criterion.

3.4.3 Oscillation with Fundamental at which the Loop Gain is Small

The numerical analysis of the previous section applies when the loop gain at the fundamental exceeds 2. This analysis was performed with the assumption that the nonlinear link is ideally memoryless. The question then arises of how fair is this widely accepted approximation of a real physical link. Would it be reasonable to assume as memoryless, for example, the effect of saturation in an amplifier if the limit frequency f_T of its transistors is 10^6 times greater than the fundamental of the oscillation?

Looking for the answer to this question, we will discuss the modes of oscillation in the feedback system shown in Fig. 3.20 [10, 113]. Here, the transistors employed in the three-stage amplifier have $f_T = 120$ MHz, and the fundamental of the oscillation is on the order of 100 Hz. The loop gain frequency response shaped by the RLC network of the feedback path is shown in Fig. 3.21.

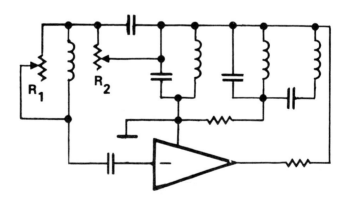

FIG. 3.20

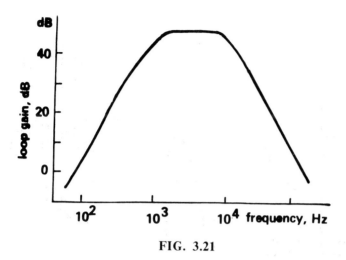

FIG. 3.21

The goal for the experiments was to determine the boundary separating the areas of conditional stability and the AGS. This boundary on the plane (y, y_f) shown in Fig. 3.22(a) relates to stopping the self-oscillation while gradually increasing the phase stability margin from 0. At sections 1, 2, and 3, this boundary was discovered by increasing y (by reducing R_1) while keeping y_f constant, and at the sections 4 and 5, conversely, by increasing y_f (by reducing R_2) while keeping y constant, as illustrated by the traces $y(y_f)$ in Fig. 3.22(a). The thicker line designates oscillation. In the area of conditional stability, the system stability and the mode of self-oscillation depend on initial conditions. In particular, the system stability and self-oscillation mode depend on the whole trace $y(y_f)$ from the point of entering the conditional stability area up to the point at which the system is examined.

Figure 3.22(b) shows the oscillation modes observed in the vicinity of the boundary. Note that the large amplitude components of the oscillation differ in frequency 10^3 times. The mode with both y and y_f comparatively large exists at the fourth section, having the fundamental of 90 Hz. The oscillation is almost sinusoidal, except in the second quarter of the period where the system bursts into oscillation with 110 kHz frequency. Let us study how these two oscillation components support each another, i.e., how does their interference in the nonlinear link create the extra phase shifts, which are positive for the fundamental ω_f and negative for the high-frequency components. These phase shifts correspondingly equal $y_f \pi$ for the fundamental and $-y\pi$ for the 110 kHz component, since the total-loop phase shift must be 2π for each of the components of a periodic oscillation.

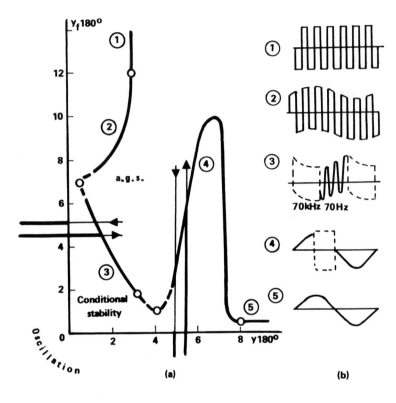

FIG. 3.22

Because of an extensive difference in the frequency of the components, the effect of the fundamental on the high-frequency components may be viewed as merely time-variable biasing for the nonlinear link (the output transistor stage) that changes its small-signal gain and phase shift. This phase shift ψ_f measured at 110 kHz under the condition of adding a sinusoidal signal with 90 Hz frequency and the amplitude E_{90} to the input of the nonlinear link is shown in Fig. 3.23. It can be seen that the phase shift is not negligible.

On the other hand, in the presence of high-frequency oscillation, the nonlinear link produces a phase lead for the fundamental. This phase lead is attributed to the nonsymmetrical limiting of the high-frequency component within the second quarter of the period, as represented in Fig. 3.24(b) by the dashed line. Let us calculate the resulting ψ_f.

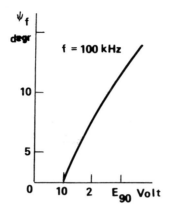

FIG. 3.23

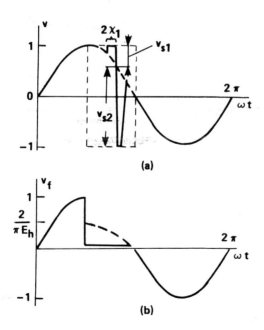

FIG. 3.24

Because $|T_0(j\omega_f)| = 1$, the amplitude of the low-frequency mode of $e(t)$ equals the threshold of saturation; we normalize this threshold to be 1. Then $v(t) = e(t)$ everywhere, except in the second quarter of the fundamental period, where the high-frequency mode appears as pictured in Fig. 3.25. E_h stands for the amplitude before limiting; χ_1 and χ_2 represent the clipping angles.

The thresholds equal

$$v_{s1} = E_h \cos \chi_1$$
$$v_{s2} = -E_h \cos \chi_2 \tag{3.15}$$

In accordance with Fig. 3.24(a)

$$v_{s1} = 1 - \cos \chi_1 t$$
$$v_{s2} = 1 + \cos \chi_2 t \tag{3.16}$$

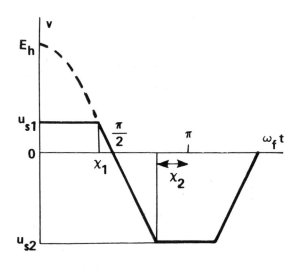

FIG. 3.25

System with a Single Nonlinear Link

In the interval $\pi/2 < \omega t < \pi$, the mean value of the high-frequency mode is

$$e_{hav} = \frac{1}{\pi} \left[\int_0^{\chi_1} v_{s1} \, d\omega t + \int_{\chi_1}^{\pi-\chi_2} E_h \cos \omega t \, d\omega t - \int_{\pi-\chi_2}^{\pi} v_{s2} \, d\omega t \right]$$

$$= \cos \omega_f \frac{1}{\pi} [v_{s1} \chi_1 - v_{s2} \chi_2 + E_h \sin \chi_2 - E_h \sin \chi_1] \quad (3.17)$$

By substituting χ_1 and χ_2 from (3.15), we obtain

$$e_h = \frac{1}{\pi} \left[v_{s1} \arccos \frac{v_{s1}}{E_h} - v_{s2} \arccos \frac{v_{s2}}{E_h} + (E_h^2 - v_{s2}^2)^{1/2} \right.$$

$$\left. - (E_h^2 - v_{s1}^2)^{1/2} + \cos \omega_f t \right] \quad (3.18)$$

The fundamental at the output of the nonlinear link

$$v_f(t) = \cos \omega_f t + v_{hav}$$

If, for example, $E_h = \infty$, then, for $\omega_f t \in [\pi/2, \pi]$, we have

$$v_f(\omega_h t) = 0.5 [v_{s1} - v_{s2}] + \cos \omega_h t = 0$$

as shown by the solid line in Fig. 3.24(b). The Fourier coefficients for the fundamental are readily found as $\operatorname{Re} V_f = 0.75$ and $\operatorname{Im} V_f = 1/2\pi$, and, therefore, the phase shift for the fundamental in the nonlinear link is $\psi_f = \arctan [\operatorname{Im} V_f / \operatorname{Re} V_f] = 0.21$. The considered mode of oscillation can, therefore, only appear when the low-frequency phase stability margin does not exceed 12°.

Smaller values of E_h lead to smaller phase shifts. This can be seen by using the first two terms of the power expansion for (3.18):

$$v_f(\omega_f t) = \frac{1}{\pi} \left[(v_{s1} - v_{s2}) \frac{\pi}{2} - \frac{1}{E_B} (v_{s1}^2 - v_{s2}^2) - \frac{V_{s2}^2}{E_B} - \frac{V_{s1}^2}{E_B} \right] +$$

$$\cos \omega_f t = \cos \omega_f t + \left[\frac{2}{\pi E_h - 1} - 1 \right] \cos \omega_f t = \frac{2}{\pi E_h} \cos \omega_h t$$

The related curve is shown in Fig. 3.24(b) by a dashed line. The calculated phase shift of the fundamental for $E_h = 2$, for example, is 7.5°.

The conclusions are that, first, the multifrequency oscillation may appear in the bandpass system if the phase stability margins are simultaneously small (less than 10°) at the lower and higher frequencies. In practice, however, the low-frequency stability margin always exceeds 30°. Thus, while designing practical bandpass systems, we may exclude the multifrequency oscillation from our consideration and confine the stability analysis to only examining the high-frequency oscillation, and execute the analysis in pretty much the same manner as for the lowpass systems.

The second conclusion is that even small deviations from the ideal characteristics of nonlinear links may affect the stability conditions quite markedly. From this point of view, the theory of absolute stability and the Liapunov direct method appear to be only approximate methods of stability analysis when we idealize the characteristics of the nonlinear elements.

3.5 DESCRIBING FUNCTION APPROACH

The ratio in complex amplitudes of the fundamental of the output $v(t)$ of the nonlinear link to the sinusoidal signal given to its input:

$$\mathbf{V}/E = H(E, j\omega) = |H(E, j\omega)| \exp[j\psi(E, j\omega)] \tag{3.19}$$

is known as the *describing function* (DF). For time-invariant links, H does not depend on the phase of the incident sinusoidal signal.

The DF stability analysis of a time-invariant feedback system containing a single nonlinear link with odd characteristic presumes that if the system oscillates, first, the self-oscillation is periodic, and, second, the shape of the oscillation at the input of the nonlinear link is nearly sinusoidal. (Note that if the characteristic is not odd, a constant component adds to the fundamental.)

With this assumption, all the harmonics in $v(t)$ are of no interest, and the system is not AGS if and only if certain E and ω exist, which satisfy the equality:

$$T_0(j\omega) H(E, j\omega) = 1 \tag{3.20}$$

To ensure that the system is stable with symmetrical rectangular margins x and $y\pi$, it is required that the locus $T_0(j\omega)$ does not encircle the critical point, and also that for the all conceivable pairs of values of E and ω the following inequality must hold:

$$|\pi - \phi - \psi| > y\pi \ : \ |20 \log |T_0 H|| < x \tag{3.21}$$

The signal $e(t)$ in a real system is not exactly sinusoidal. The mutual interference of its harmonics in the nonlinear element contributes to the fundamental at its output. However, the changes in the value of the fundamental are small in most instances because of the following two circumstances [157]:

(a) $T_0(j\omega)$ possesses the properties of a filter, attenuating the harmonics in the return signal;
(b) the characteristic $v(s)$ is such that the harmonics of small amplitudes still contained in e have almost no effect on the fundamental of v.

The assumption (b) is quite fair for the ordinary nonlinear links of control systems. Condition (a) is fulfilled for the optimally designed systems (recall Fig. I.1 of the Introduction), as discussed below.

In a narrowband resonance system without inversion in the feedback loop, it is natural to suppose that the fundamental of the oscillation belongs to the passband of the filter (*conjecture of a resonance*); then, $T_0(j\omega)$ filters the harmonics off the signal fed back. With this, the oscillation parameters (frequency, amplitude, *et cetera*) can be calculated with very reasonable accuracy.

The *conjecture of a filter* is applied to the design of stable lowpass control systems, mainly those with small NPS and comprising nondynamic nonlinear links [57, 165]. The conjecture asserts that if a periodic self-oscillation exists in such a system, the fundamental ω_f falls in the cut-off frequency range of the loop gain response, beyond the range of flat gain, because, due to the Bode phase-gain functional, the phase condition of the oscillation is $\arg T_0(j\omega_f) \cong -\pi$, which may only occur on the sloped part of the Bode diagram, or nearby on its left-hand side. The average steepness of the diagram is $-12\,\text{dB/octave}$, so that the third harmonic component returning to the input of the nonlinear link is damped by $12\log_2 3 \cong 20\,\text{dB}$ relative to the fundamental. Taking into account that, in the most common Π-shaped type of $v(t)$, the third harmonic is already 10 dB down from the fundamental, we conclude that in $e(t)$ the amplitude of the third harmonic is usually 30 dB (i.e., 30 times) smaller than the fundamental. The higher harmonics are damped even more, so that condition (a) is guaranteed.

Because the harmonics in $e(t)$ are small, their interference could create a phase shift for the fundamental of only 5° to 15°. The accuracy of the stability margin estimation with the DF is, therefore, high. The accuracy in determination of the frequency of oscillation is poor, especially in systems having small stability margins over a broad frequency range, but this is not of importance for control system design.

The errors of the DF method can be evaluated with computer simulation, employing Galerkin's method, or using special integral relationships [157]. The accuracy of the solution can be improved by taking into consideration the higher harmonics in $e(t)$. This may be required when studying the systems with smaller selectivity of the feedback loop, i.e., systems including nonlinear links that produce large phase lag for the fundamental, such as links with large hysteresis or nonlinear dynamic links, and also systems with a resonant peak on the $|T(j\omega)|$ response.

Checking the local stability of the solution completes the analysis of the problem solved with the DF technique [5]. When the solution is single-valued, it is stable as a rule, and the local stability analysis may be omitted.

The DF analysis, therefore, replaces the nonlinear link by an equivalent linear link with the transmission function $H(j\omega)$. The only difference is that H depends on E, and hence neither the superposition principle nor the Bode integral relationships remain applicable for the frequency responses measured with the constant amplitude of the signal generator. When the source of signal amplitude varies with frequency in accordance with a specified low, the features of the frequency responses of the DF become closer to those of the linear circuit functions, and the Bode relations may be applied, as will be shown in Chapter 5.

3.6 DESCRIBING FUNCTIONS FOR THE BASIC TYPES OF NONLINEAR LINKS

Hysteresis. In a link with hysteresis-like output-input dependence, shown in Fig. 3.26, the output v depends on the prehistory of the input e, and must be found by using the branch c_1 within the time-intervals of rising $e(t)$, and the branch c_2 when $e(t)$ is decreasing.

From the definition of DF:

$$\text{Re } H = \frac{1}{\pi E} \int_0^{2\pi} v \sin \omega t \, d\omega t$$

$$\text{Im } H = \frac{1}{\pi E} \int_0^{2\pi} v \cos \omega t \, d\omega t$$

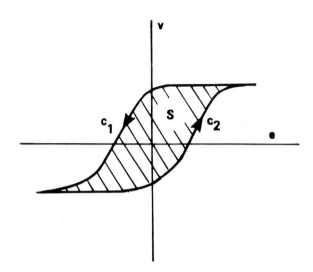

FIG. 3.26

it follows that

$$\text{Im } H = \frac{1}{\pi E} \int_0^{2\pi} v \, d \sin \omega t = \frac{1}{E^2} \oint_{c_1 + c_2} v \, de = \frac{S}{\pi E^2} \tag{3.22}$$

where S stands for the area within the hysteresis loop. Then,

$$\arg H = \arcsin \frac{\text{Im } H}{H} = \arcsin \frac{S}{\pi E^2} \frac{E}{V} = \arcsin \frac{S}{\pi EV} \tag{3.23}$$

Because $|\arg H|$ is small in most well-designed closed-loop systems, we will pay no special attention to the hysteresis in the following chapters. If necessary, the phase lag generated by the hysteresis can be compensated by a nonlinear dynamic compensator (to be studied in Chapters 5, 7, and 8).

Using logic circuits, a nonlinear link could be designed to emulate negative hysteresis for specified types of periodic input signals. Such links improve stability margins for the signals of certain frequencies and amplitudes, but also simultaneously reduce the stability margins for the signals in other ranges of frequency and amplitude.

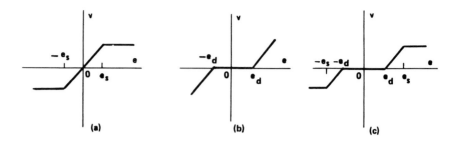

FIG. 3.27

Saturation. For nonlinear links with symmetrical hard saturation, having the threshold e_s and unity gain, as shown by the solid line in Fig. 3.27(a), the DF is given for $E \geq e_s$ [168] by

$$H = \frac{2}{\pi} \arcsin \frac{e_s}{\pi E} + \frac{e_s}{\pi E} \sqrt{1 - \frac{e_s^2}{\pi^2 E^2}} \qquad (3.24)$$

However, this formula is inconvenient for the synthesis of a system that includes several nonlinear links. Instead, we will use its approximation [162], modified in [113] as

$$H \cong 1.27 \, (E/e_s)^{-1} - 0.27 \, (E/e_s)^{-4} \qquad (3.25)$$

which has an error of less than 1%, i.e., smaller than 0.1 dB. This is less than negligible as compared with the tolerances of the linear link's gain and the error of approximation of the real link's nonlinearity by hard saturation. In this formula, the second component correspondingly contributes only 2 dB, 0.6 dB, and 0.35 dB for the normalized signal amplitude E/e_s equal to 1, 1.5, and 2. This component may be omitted for $E/e_s > 1.5$.

In developing synthesis procedures for feedback systems, we will deal mostly with actuators having a saturation characteristic. Saturation provides the attractive feature of maximal and constant gain up to the maximum of the output amplitude, in contrast to other types of nonlinear links. Hence, when the loop gain is limited by stability conditions, the actuator transfer characteristic is best when close to saturation. When it is not, a predistortion memoryless nonlinear link could be installed at the input to the actuator to make the total nonlinearity the saturation.

Dead zone. The link with the dead zone characteristic of Fig. 3.27(b) can be equivalently presented as the parallel connection of an inverting link with saturation, having the threshold e_d, and a unity link. The fundamentals of the outputs of the links added together give the fundamental for the total link. Hence, the DF for the total link is the difference of the DF values for the elementary links. Then, for $E \geq e_d$:

$$H \cong 1 - 1.27\, e_d/E + 0.27\, e_d^4/E^4 \tag{3.26}$$

If the nonlinear link characteristics include both dead zone and saturation as shown in Fig. 3.26(c), then, similarly for $E \geq e_s$:

$$H \cong 1.27\,(e_s - e_d)/E - 0.27\,(e_s^4 - e_d^4)/E^4 \tag{3.27}$$

For the input signals of large amplitudes, the second component in (3.26) and (3.27) may be omitted.

Local feedback path. The linear local feedback path makes the differential gain of the nonlinear link almost constant over the amplitude range where the feedback is large, i.e., over the range of amplitudes bounded by the maximal swing of the output that can be delivered. Hence, the feedback makes the nonlinear characteristic curve of the output-to-input dependence more linear within the working range of signal amplitudes, sharpens the previously rounded corners to form thresholds, and reduces the dead zone (if any exists). The resulting link characteristic approximates ideal saturation, which, as mentioned, benefits system performance. Such local feedback is widely employed in electrical and electromechanical actuators.

Nondynamic saturation with a frequency-dependent threshold. Such a link can be realized, for instance, as shown in Fig. 3.28. Replacing e_s in (3.25) by $e_s k(j\omega)$ gives the DF of the link. A link with a frequency-dependent width of the dead zone can be made in a similar manner.

Dynamic saturation. The link often encountered in practice is one whose $|H|$ is the same as that of memoryless saturation, but $\arg H = \psi(\omega, E) \neq 0$, i.e.,

$$H \cong (1.27\, e_s/E - 0.27\, e_s^4/E^4)\exp j\psi(\omega, E) \tag{3.28}$$

We will call this link *dynamic saturation*.

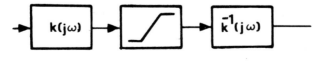

FIG. 3.28

A typical case is presented in Fig. 3.29, where a link with memoryless saturation (final gain stage) follows a dynamic nonlinear link (prefinal stage), which is fairly linear within the range of linearity of the final link, so that $|H|$ is determined by the final stage. The phase shift arg H could be substantial if the prefinal stage comprises local feedback, reactive two-poles shunted by diodes, nonlinear dynamic compensators *et cetera*.

Figure 3.30 presents two more examples of dynamic saturation. In Fig. 3.30(a), the high-gain linear link $\mu(j\omega)$ and the link with saturation are looped by the linear feedback path β_l. The module of the output fundamental is determined by the saturation link, and its changes with the level of the input. For small amplitudes, $\psi = 0$; for large amplitudes, the DF of the saturation link decreases and the local feedback vanishes, making $\psi \cong \arg(1 + \mu\beta_l)$.

The system shown in Fig. 3.30(b), consisting of a link looped with a feedback path that comprises a dead zone link, will be analysed in more detail in Chapter 6.

The *Clegg integrator* shown in Fig. 3.31 consists of a splitter, two different linear links L_1 and L_2, a full-wave rectifier, a high-gain link with saturation realizing the operator sign $v_1(t)$, and a multiplier M [37]. If $L_1 = L_2$, the whole composed link behaves as L_1. Otherwise, the module of the output signal is determined by the upper channel output $v_1(t)$, and the sign of the output is determined by the lower channel, so that $v(t) = |v_1(t)|$ sign $v_2(t)$.

Employing, for example, an integrator as L_1, and $L_2 = 1$, the loop gain decreases with frequency without introducing the conventional linear-integrator phase lag of $90°$, as shown in Fig. 3.32. However, the high sensitivity of H to the shape of the input signal (also, the disadvantage of active hysteresis) prevents this circuit from being widely used.

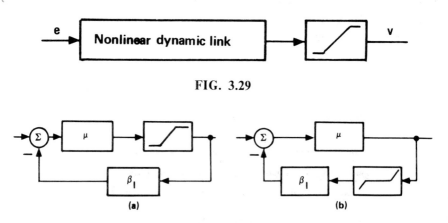

FIG. 3.29

FIG. 3.30

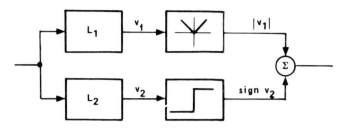

FIG. 3.31

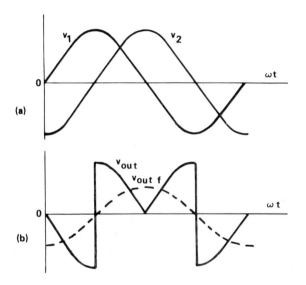

FIG. 3.32

3.7 MULTIVALUED OUTPUT-INPUT RELATIONS

3.7.1 Two-Pole with Negative dc Resistance

Liapunov's first method claims that the local stability conditions for a nonlinear differentiable system follow the stability condition for a locally linearized system [33, 42]. The latter can be investigated using Nyquist diagrams, Rouchet theorem [23, 145] *et cetera*. For certain nonlinear systems, however, analysis of linearized systems may be avoided, and a judgement can be made about system stability on the basis of the types of nonlinear functions involved. This approach, based on considerations of continuity and topology, yields the interrelations between certain types of nonlinearities and the type of frequency response of the linearized system [96, 30, 31, 113].

Consider first a two-pole with the type of dc current-voltage dependence illustrated in Fig. 3.33. Suppose that the linearized circuit is adequately described by a system of a finite number of linear differential equations, and the coefficients of these equations continuously depend on the dc biasing. In such systems, the zeros and poles of the impedance $Z(s)$ of the locally linearized circuit migrate along certain continuous curves (root loci) in the s-plane as the biasing point moves continuously along the current-voltage characteristic curve.

Next, we presume that due to the stochastic character of real world systems, only a zero probability exists of crossing the $j\omega$-axis simultaneously (i.e., at the same biasing point) by several independent root loci, and exclude such an event from the analysis. Then, it is considered that at any biasing point the impedance $Z(s)$ may only have on the $j\omega$-axis either a single real zero (at the origin), a single real pole (at the origin), a pair of imaginary conjugate zeros, or a pair of imaginary conjugate poles. Correspondingly, $Z(s)$ approaches either 0 or ∞ at the crossover frequency.

It follows, in particular, that at the instant when the number of positive real zeros changes, the resistance $r_0 = Z(s) : s = 0$ reaches 0, and when the number of positive real poles of Z changes, r_0 becomes ∞. In the example of Fig. 3.33, this situation correspondingly occurs at points A_1, A_2, and B_1, B_2.

We also assume that some physical line connects the external terminals of the two-pole to its internal structure, whereby the attenuation of the line increases infinitely with frequency (due to the skin effect, *et cetera*). At infinite frequency, the impedance Z, therefore, equals the wave impedance r_c of the line, and is not dependent on the biasing.*

In accordance with the argument principle, the difference between the number of zeros and the number of poles of $Z(s)$ in the right half-plane of s equals the number of turns by π on the frequency hodograph $Z(j\omega)$, which are counted while ω changes from 0 to ∞. (If Z becomes infinite at a certain frequency, the hodograph must be completed with arcs of infinite radius to close the Nyquist contour [23].) If sign $r_0 = -\text{sign}\, r_c$, this difference is odd; hence, the difference between the numbers of real positive poles and real positive zeros is also odd. (Note that the number of complex poles and zeros is always even, so it does not affect the calculation.)

---

*In effect, the powers of the polynomials of the numerator and denominator of $Z(s)$ are assumed to be equal. We can also arrive at the results obtained in this section using weaker assertions for these powers differing by a maximum of 1, and $\lim(s \to \infty) Z(s)$ is not dependent on the biasing.

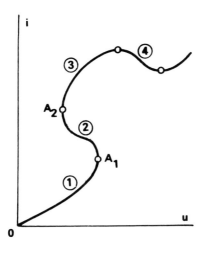

FIG. 3.33

Therefore, if the two-pole at the first section of the dc dependence in Fig. 3.33 is stable while it is both unloaded and shorted, i.e., $Z(s)$ possesses neither zeros nor poles in the right half-plane of s, then $Z(s)$ has an odd number of positive real zeros immediately beyond the point A_1. Neglecting, as assumed, the zero probability cases of several real zeros or poles simultaneously crossing the origin, we conclude that $Z(s)$ possesses one positive zero at the second section of the dc characteristic curve.

It is interesting to discover that at small ω, the impedance $Z(j\omega)$ is inductive in the neighborhood of point A_1, i.e., $\operatorname{Im} Z(j\omega) > 0$, since in this only case the number of turns by π on the hodograph for $Z(j\omega)$ may increase by 1, while the biasing point passes the point A_1 and, therefore, r_0 passes through 0.

A similar argumentation shows that $1/\operatorname{Im}[1/Z(j\omega)] > 0$ in the vicinity of the point B_1 for small enough ω; and at the fourth part of the dc characteristic, the function $Z(s)$ possesses a positive real pole so that the two-pole is unstable when unloaded.

Coupling the considered two-pole to a linear passive one and applying the same kind of reasoning to the resulting composite two-pole, we arrive at the following statement: the circuit is unstable if the biasing point belongs to the fourth section and the dc resistance of the passive two-pole is larger than r_0; or, if the biasing point belongs to the second section, and the dc resistance of the passive two-pole is smaller than r_0.

Assuming that at the first section $r_0 > 1$ and the two-pole is stable being either open or shorted, we will refer to *dropping sections* as those negative resistance sections, between which and the first section, the number of points

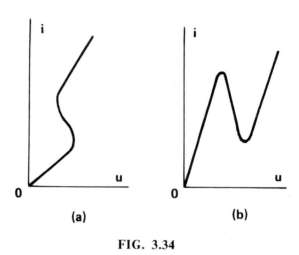

FIG. 3.34

of zero derivative, counted along the dc characteristic, is odd. On the dropping sections, $Z(s)$ possesses a real positive zero that makes the two-pole unstable if shorted.

In the usual case, the dc curve does not intersect itself, and either one or three values of the current relay to the given voltage. The negative resistance sections of such a curve may be of two types only: S or N, as illustrated in Fig. 3.34.

3.7.2 Two-Port

Suppose the curve plotted in Fig. 3.33 represents the dc output-input dependence of a two-port. Assume that at the first section of the curve the system is locally stable. Hence, the main determinant of the locally linearized system (which constitutes the denominator of the output-input transfer ratio), in particular, does not possess positive-real zeros. Then, at the dropping sections, the main determinant has positive-real zero and the system is unstable.

We have previously studied the multivalued dependencies of dc signals. This consideration can be extended to the analysis of systems with multivalued dependencies of either fundamental, harmonics, or subharmonics on either the amplitude, frequency, or some other parameter of the input signal.

Consider, as an example, the connection of the three links shown in Fig. 3.35. The modulator M (an amplitude modulator, or a voltage-controlled oscillator) generates a periodic signal whose parameters (amplitude, frequency, or shape) are determined by the input signal U. The demodulator

(DM) measures the assigned parameter of the output signal of the nonlinear link N (amplitude, phase, frequency, amplitude of a certain harmonic or subharmonic, the mean value, the mean square value, *et cetera*) producing the dc current I proportional to the measured parameter.

Assume that M and DM are locally stable for all conceivable signals (as are the commonly used passive devices) so that the transfer function of the locally linearized tandem connection M-DM does not possess positive-real zeros, and the dependence $I(U)$ is single-valued.

Suppose now that the two-port N is locally stable at $U=0$, and the dependence $I(U)$ of the system M-N-DM has a dropping section somewhere. Then, at this section, the linearized transfer coefficient, according to the previous analysis, possesses a positive-real zero. Hence, the signal corresponding to this Laplace transform has an exponential term with the exponent βt, where β is real and positive. Since the links M and DM are locally stable, the exponential term certainly appears due to the two-port N. Hence, βt is the characteristic exponent [42] of the output of the linearized (time variable) two-port N, and for the two-port N as well.

It follows that over the interval of U where $I(U)$ drops, the dependence of any other parameter of the link's N output signal must also have a dropping section, although the values of the related jumps can be quite different. For example, while the amplitude of the fundamental jumps down due to the jump-resonance phenomenon (to be described in the next chapter), the phase of the fundamental also jumps, and the amplitudes of the harmonics jump down from some finite values to nearly 0.

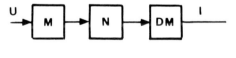

FIG. 3.35

Chapter 4
Forced Oscillation

4.1 PERIODIC EXCITATION

In response to periodic excitation, the output signal of a nonlinear feedback system may or may not be periodic. We will concentrate our study on periodic responses, since the nonperiodic responses observed in the systems of practical interest, see Fig. I.1, have their spectral densities localized within narrow frequency bands and submit to study by the same methods as those applied to periodic signals.

The amplitude and the shape of a periodic output signal generally depend on the input signal's prehistory, and they may be multivalued functions of the input signal parameter. Particularly, the output might depend on whether the current value of the input signal was arrived at by gradually increasing or decreasing the input signal amplitude.

This phenomenon may be observed in the nonlinear feedback system of Fig. 4.1(a), excited by a sinusoidal input $u = U\sin\omega t$. Its nature is, however, easier to understand by initially studying the inverted system of Fig. 4.1(b), which implements the same relationships (as demonstrated in Sec. 1.1.3).

In order to use DF approach, we assume that $e(t) = E\sin\omega t$. The legitimacy of this approximation will be discussed in Sec. 4.3.

When the phase stability margin is rather small, let us say, 30°, i.e., arg T_0 is $-150°$, and the nonlinear link is memoryless so that arg $H = 0$, then the mutual compensation of the outputs of the two parallel channels generates a collapse on the characteristic $U(E)$ in the region where these outputs are nearly equal in value. This is illustrated in Fig. 4.2(a) for a system with saturation, and in Fig. 4.2(b) for a system with dead zone.

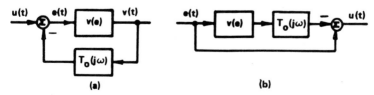

FIG. 4.1

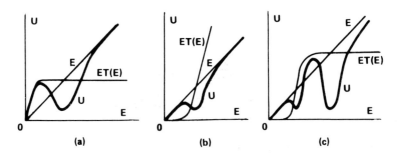

FIG. 4.2

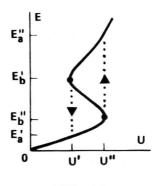

FIG. 4.3

The output-input dependence $U(E)$ for the system of Fig. 4.1(b) is single-valued. Therefore, it may be measured experimentally. Conversely, for the system of Fig. 4.1(a), the same curve represents the dependence of the input on the output. The inverse operator, the dependence of the output on the input, is triple-valued. This dependence, redrawn in Fig. 4.3, comprises a branch dropping over the interval (U', U'') between the points of bifurcation, marked by the bold dots. As previously shown in Sec. 3.7, the solutions to the system's equations for this branch are unstable, causing the output signal to jump between the two stable solutions.

An upward jump takes place when U is gradually raised to U'' from an initial value of less than U', and a downward jump takes place when U is reduced to U' from a certain value exceeding U''. This phenomenon was given the name of *jump-resonance* when first studied in nonlinear resonance contours, where it manifests itself in jumps on the frequency response curve in the vicinity of the peak of resonance.

Forced Oscillation

As shown, jump resonance is observed in feedback systems when arg T is such that the function $U(E)$ possesses a minimum. Jump resonance happens within certain frequency intervals. At the ends of these intervals, the limit case occurs of $U' = U''$ and $dU/dE = 0$, as shown in Fig. 4.4.

Figure 4.5 shows a family of curves $E(U)$ in some frequency range. The bifurcation points constitute a closed space curve R. Its projection r upon the plane (ω, U) encloses the area of jump-resonance existence.

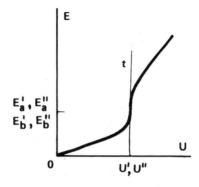

FIG. 4.4

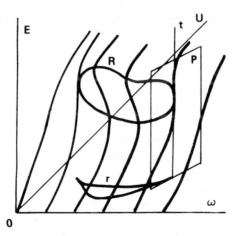

FIG. 4.5

For the physical systems of interest, it is reasonable to suppose the curve R is of finite curvature. A tangent line can be drawn to any point of such curve. In the limit case, the tangent t is perpendicular to the plane (ω, U). Through t, a plane P is drawn tangent to R as shown in Fig. 4.5. Since the part of R adjacent to P lies to one side of t, its projection forms a pointing on the curve r, with the limit case producing the tip of the pointing.

The number of the pointings is even. In the usual case, this number is two, so the pointings break r into two branches, the first associated with U'', and the second with U'.

If U and ω are gradually varied along some continuous curve c crossing r, a jump occurs at the point of leaving the interior of r, when the number of pointings between this point and the point where c first entered the interior of r is odd. In Fig. 4.6, the crosspoint associated with the jump is marked by a bold dot.

In this chapter, we confine our analysis to the most frequently encountered triple-valued functions $E(U)$, although the analysis may be readily extended to treat more complicated curves r and dependencies $E(U)$, such as, for example, the dependencies comprising two dropping branches illustrated in Fig. 4.2(c). Such curves may be generated by a feedback system with the nonlinear link possessing both a dead zone and saturation. An example of such a link will be furnished in Sec. 7.2.3.

When a frequency response is measured with the constant amplitude of the input excitation, the curve c is parallel to the ω-axis. In a nonlinear system with a second-order linear part described by Duffing's equation [26, 63]:

$$\frac{d^2 e}{dt^2} + \frac{de}{dt} + v(e) = U \cos \omega t$$

where $v(e)$ is either smooth saturation or smooth dead zone, the type of this nonlinear characteristic uniquely determines the directions of the jumps in the frequency response. With increasing frequency, E jumps up in systems with saturation, and down in systems with dead zone. With decreasing frequency, E jumps in reverse order.

This simple rule fails for systems with a higher order of linear part, where the upward and downward jumps may coexist with the nonlinear element of either type. This can be readily seen in the example of Fig. 4.7, where a linear filter at the output of a signal source with constant signal amplitude U_1 forms the curve c. Provided that the curve c alternately crosses the upper and lower branches of r several times, several jumps occur. For the case illustrated in Fig. 4.8(a,b), first, a downward jump, and then an upward jump take place.

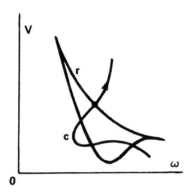

FIG. 4.6

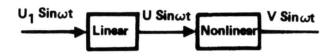

FIG. 4.7

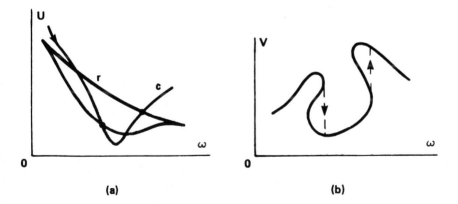

FIG. 4.8

It is instructive to mention that, generally, the frequency response measurements may not disclose all the regions where the output-input dependence is multivalued. Consider, for example, the region of jump-resonance and the frequency responses of the experimental device [113] shown in Fig. 4.9. The isolated branches of the responses, even the upper branch related to the stable solution, cannot be observed when the line c is parallel to the ω-axis. The isolated branches were discovered by increasing the input amplitude and then gradually reducing it, such that c successively crossed the upper and lower branches of the curve r.

Jump-resonance will be studied quantitatively in Sec. 4.3.

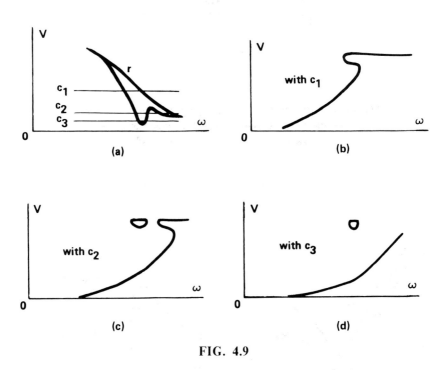

FIG. 4.9

4.2 ABSOLUTE STABILITY OF THE OUTPUT PROCESS

The output process is considered unstable if an infinitesimal deviation in the input causes a finite increment of the output. Particularly, the process related to the inner branch of the triple-valued dependence $E(U)$ is unstable.

A system is only said to be process-stable when the output processes are stable for all conceivable inputs. It follows, in particular, that a system having multivalued dependence $E(U)$, as described in the previous section, is not process-stable.

Process stability implies the stability of the locally linearized (generally, time variable) system.

The system of Fig. 4.1(a) is said to be absolutely process-stable (APS) if it is process-stable with any characteristic of the nonlinear memoryless link $v(e)$ whose derivative is limited by

$$0 < \frac{dv}{de} < 1 \qquad (4.1)$$

The system is APS [159, 170] if at all frequencies:

$$\text{Re } T_0(j\omega) > -1 \qquad (4.2)$$

The useful equivalent forms of this condition are

$$\text{Re } F_0(j\omega) > 0 \qquad (4.3)$$

$$\text{Re } 1/F_0(j\omega) > 0 \qquad (4.4)$$

$$\cos \arg T_0(j\omega) > 1/|T_0(j\omega)| \qquad (4.5)$$

and

$$|M_0(j\omega)| = \left|\frac{T_0(j\omega)}{F_0(j\omega)}\right| > \frac{1}{|\sin \arg T_0(j\omega)|} \qquad (4.6)$$

The condition (4.3) is illustrated as follows. The feedback system linearized for deviations is shown in Fig. 4.10(a), where

$$g(t) = \frac{dv}{de} \in (0,1)$$

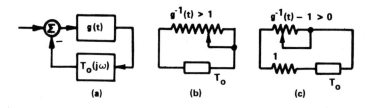

FIG. 4.10

in accordance with (4.1). Exploiting the analogy between the feedback system and the two-pole connection discussed in Sec. 1.1, the equivalent circuits are shown in Fig. 4.10(b,c). The function $F_0(s) = T_0(s) + 1$ is positive-real, so the lower branch of the circuit in Fig. 4.10(c) is physically realizable as a passive time-invariable two-pole. The only time-variable part of the system, $1/g(t)$, is resistive and positive, i.e., energy-dissipative. Hence, the oscillation cannot arise in such a system.

The stability of the linearized system follows from the foregoing discussion, despite the fact that g may be time-variable, whether due to modulation of the nonlinear link $v(e)$ by the signal, or to direct variations of the link parameter with time. Therefore, the inequalities (4.2) to (4.6) guarantee the output process stability in feedback systems with time-dependent, memoryless links of the kind that satisfy (4.1).

The conditions (4.2) to (4.6) are not only sufficient, but also necessary, for the APS. To prove this, we intend to show that if $\text{Re } T_0 < -1$ at some frequency ω, then there exist (a) some function $v(e)$ satisfying (4.1) and (b) some periodical input signal with the fundamental ω, which together produce an unstable output process [110, 113].

Proof. We confine ourselves to a symmetrical single-valued, twice continuously differentiable function $v(e)$, and a symmetrical periodic wave $e(t)$. Consequently, the wave $v(t)$ is symmetrical as well, and the associated Fourier series contain sinusoidal components only. Then, the relation between the amplitudes U_k, E_k, V_k of kth harmonics of $u(t)$, $e(t)$, and $v(t)$, correspondingly, can be written as

$$U_k^2 = [E_k + V_k \text{ Re } T_0 (j k \omega)]^2 + [V_k \text{ Im } T_0 (j k \omega)]^2$$
$$= E_k + 2 E_k V_k \text{ Re } T_0 (j k \omega) + V_k^2 | T_0 (j k \omega) |^2$$

From here, the dependence $E_k (U_k)$ can be found.

If there exists a value of E_k for which

$$\frac{d U_k}{d E_k} < 0$$

the dependence $E_k (U_k)$ is certainly multivalued.

Because $U_k > 0$, by multiplying by $2 U_k$, we obtain an equivalent inequality:

$$\frac{d U_k^2}{d E_k} < 0$$

Forced Oscillation

By substituting U_k^2, we obtain

$$E_k + V_k \operatorname{Re} T_0(j\omega) + E_k \frac{dV_k}{dE_k} \operatorname{Re} T_0(jk\omega) + |T_0(jk\omega)|^2 V_k \frac{dV_k}{dE_k} < 0$$

so that

$$\operatorname{Re} T_0(jk\omega) < -\frac{1 + |T_0(jk\omega)|(V_k/E_k)(dV_k/dE_k)}{V_k/E_k + dV_k/dE_k} \tag{4.7}$$

Next, let the periodic input signal $u(t)$ be such that

$$e = E_0 \operatorname{sign} \sin \omega t$$

where E_0 is the amplitude of these pulses. Because the characteristic $v(e)$ was assumed to be symmetrical, $v(t)$ is also a sequence of symmetrical pulses, i.e.,

$$v(t) = m E_0 \operatorname{sign} \sin \omega t$$

with m being the ratio of the output and input amplitudes of the nonlinear link. Because, further, for any k, the equality $V_k/E_k = m$ holds, and

$$\frac{dv}{de} = \frac{dV_k}{dE_k}$$

we rewrite (4.7) in the form

$$\operatorname{Re} T_0(jk\omega) < -\frac{1 + m|T_0(jk\omega)|^2 dv/de}{m + dv/de} : e = E_0 \tag{4.8}$$

Particularly, if $dv/de = 0$, then $\operatorname{Re} T_0(jk\omega) = -1/m$. Further, if the operator $v(e)$ represents smooth saturation that closely approximates hard saturation, the amplitude $E_0 = 1 + 0$ leads to $m = 1$; and, for the fundamental ($k = 1$), (4.8) yields the condition:

$$\operatorname{Re} T(j\omega) < -1 \tag{4.9}$$

Q.E.D.

The limit derivative approximation of the dead zone shown in Fig. 3.27 serves as another example. If, again, $E_0 = 1 + 0$, then $m = 0$, $dv/de = 1$, and (4.4) applied to the fundamental again yields the condition (4.9).

Corollary. If the system with saturation (or with dead zone) is process-stable, it is absolutely process stable, i.e., it remains process-stable after replacing the saturation (or dead zone) by any other nondynamic nonlinear link possessing a twice-differentiable characteristic satisfying the limitation (4.1).

The condition (4.3) plotted on the Nichols chart in Fig. 4.11 forms the boundary for the area where forced oscillation may take place.

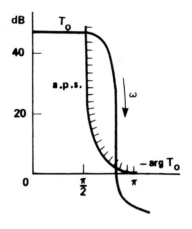

FIG. 4.11

Next, let us show in another fashion that the boundaries of forced oscillation must coincide for systems with saturation and those with dead zone. First, by transporting the input in the linear part, we transform the system of Fig. 4.10 containing a dead zone element into the system of Fig. 4.12(a). Certainly, conditions for jump-resonance are the same for both these systems. Further, by simultaneously changing the signs of the transmission coefficients of the two links in Fig. 4.12(a) and adding two unitary branches that compensate each other, the equivalent block diagram of Fig. 4.12(b) results. Finally, the equivalent transformation yields the block diagram of Fig. 4.12(c). This system has a saturation element and a linear part with the transfer function $T_0' = 1/(T_0 + 1) - 1$. Since

$$F_0' = 1/F_0 \qquad (4.10)$$

condition (4.3) applied to F_0 secures the inequality (4.4) with respect to F_0', and *vice versa*.

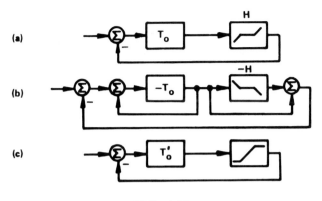

FIG. 4.12

The APS system is totally free from all forms of forced oscillation, including jump-resonance and subharmonics. However, in such systems the inequalities (4.2) and (4.4) strongly restrict arg T_0, and thereby severely constrain the maximum available feedback. That is why the APS conditions are commonly abated in control systems and feedback amplifiers. As illustrated in Fig. 4.11, the typical hodograph for T_0 of a practical system crosses the border of the APS area and, for some input signals, the output process becomes unstable.

The statistical error in the system output caused by process instability depends on the typical shape of the input signals. For example, the system process instability does not show up in the output nonlinear noise if the input signal statistic reminds us of that of white noise [12]. On the other hand, if the input signal includes high-level bursts for several periods of oscillation, the error due to process instability could be substantial.

The large amplitude step-function inputs applied to the non-APS system may produce output transient responses having large and long overshoots. This is illustrated in Fig. 4.13(b) for the system of Fig. 4.13(a), the Bode diagram of which is shown in Fig. 4.13(c). This phenomenon, called *windup*, can be crudely understood as in the following description.

The duration of the overshoot is equal to a half-period of a certain frequency at which $|T_0|$ is large, $y\pi$ small, and, therefore, the process instability is strong. The interrelations between the small-amplitude and large-amplitude frequency and transient responses are symbolized by the arrows in Fig. 4.13(b,c). Loosely speaking, due to saturation and related reduction of the loop gain, the Bode diagram for T_0 "sinks," as shown by the dotted line, and crosses the 0 dB line in the region where the phase stability margin is relatively small, thereby producing a strong overshoot in the transient response. These explanations, although far from rigorous, help to

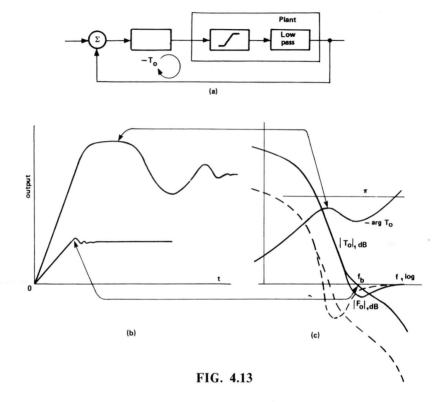

FIG. 4.13

understand the wind-up phenomenon and show possible ways in which it may be reduced or eliminated.

We conclude that the APS systems are not practical because of reduced feedback, but, on the other hand, the systems with too great a process instability are also not desirable. The required feedback system performance should, therefore, lie somewhere between these two extremes.

To evaluate and certify the systems, certain test signals should be selected. When using sinusoidal inputs, the values of the jumps of jump-resonance may be employed as the figure of merit.

4.3 JUMP-RESONANCE

4.3.1 Conditions for the Jumps

The jump-resonance phenomenon, already clarified in Sec. 4.1, can be examined via various techniques. In the early literature, the Duffing and the third-order differential equations were investigated directly [26, 63]. Application of the DF approach initiated by M.A. Aizerman [4] provided a powerful tool for the analysis of jump-resonance in feedback systems with a high-order linear part. For the system with saturation, the boundary of the domain on the T_0-plane where jump-resonance exists was obtained as the envelope of a family of circles. Similar results have been obtained by Levinson [90]. These methods have been further developed in other references [27, 62, 41, 49, 50, 113, 150, 160]. The isolated branches of a frequency response were discovered by Kozlowski [86].

Within the dynamic range of practical interest in actual systems, the dependence $E(U)$ obeys, as a rule, the linear low for both sufficiently small and sufficiently large input amplitudes. Then, as Fig. 4.14 indicates, the number of solutions corresponding to any given U is odd. Each solution of an even number is unstable, as follows from the assertion of Sec. 3.7.

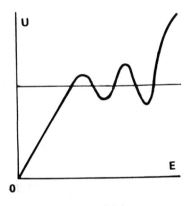

FIG. 4.14

Let us examine a system with a relatively simple nonlinear link characteristic, causing only one region of jumps, with only three solutions, as is charted in Fig. 4.15. The amplitudes at the bifurcation points, just before the jumps, E'_b and E''_b, are the roots of the equation:

$$\frac{dU}{dE} = 0$$

and, since U is positive, also those of the equation:

$$\frac{dU^2}{dE} = 0 \tag{4.11}$$

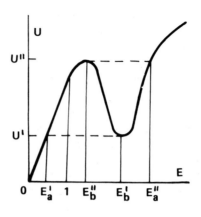

FIG. 4.15

(Note that dealing with squared values simplifies the analysis.) Substituting $U^2 = E^2|F^2(E)|$ into (4.11) gives

$$E^2 \frac{d|F^2|}{dE} + 2E|F^2| = 0 \qquad (4.12)$$

After permutation of

$$|F^2| = |T_0 H(E) + 1|^2 = 1 + 2|T_0|H\cos\arg(T_0 H) + |T_0|^2 H^2 \qquad (4.13)$$

into (4.12), E_b' and E_b'' can be found and plotted onto the T_0-plane.

When the system's parameters are at the border of the area in the T_0-plane where the jumps exist, U' meets U'', and E_b' and E_b'' become confluent, forming a double root E_b, as seen in Fig. 4.16. This root is that of the equation obtained by derivation of (4.12):

$$E^2 \frac{d^2|F^2|}{dE^2} + 4E \frac{d|F^2|}{dE} + 2|F^2| = 0 \qquad (4.14)$$

Permutation of (4.13) into (4.14) yields a relationship between $|T_0|$, $\arg T_0$, and E_b. An equation determining the boundary of jump-resonance on the T_0-plane can be deduced by studying under what condition E_b is real and positive.

Because the hysteresis reduces the available feedback, it has to be excluded or reduced to be negligible in order to maximize the feedback. For this

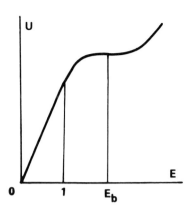

FIG. 4.16

reason, we will contemplate the jump-resonance phenomenon in systems with single-valued dependences $v(e)$ (saturation, dead zone, and power function) and dynamic saturation. The effect of small hysteresis on the value of jumps can be taken into account by correspondingly increasing the loop phase lag.

4.3.2 System with Dynamic Saturation

Determine the maximum values of the jumps in V caused by the jumps in E.

Because the jumps cannot exist in a linear system, in a system with saturation the bifurcation point coordinate E_b'' must exceed the threshold of saturation, normalized to be 1, i.e., $E_b'' > 1$. Therefore, the related amplitude of the fundamental at the output of the nonlinear link is $V_b'' > 1$ as well. In the process of jumping up, E_a'' replaces E_b'' and creates a new value of the fundamental at the system's output V_a'' that is larger than the previous value of V_b''. From (3.23), this value is certainly less than 1.27, as illustrated in Figs. 4.17 and 4.18. Consequently, $V_a''/V_b'' < 1.27$. In real systems having a fairly large phase stability margin, this ratio is even smaller, less than 1.1. Such small jumps are usually viewed as acceptable, especially because they only appear under the circumstances of overloading the actuator, when the error is already large.

In the process of jumping down, E decreases from $E_b' > 1$ to E_a'. If $E_a' < 1$, the jump in V may be large, as shown in Fig. 4.18, and may cause a large error at the output. From practical experience (and the analysis of Sec. 4.2), feedback systems with large jumps of jump-resonance are also systems with

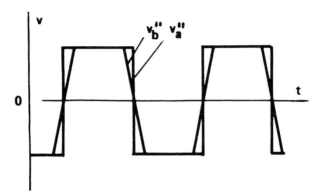

FIG. 4.17

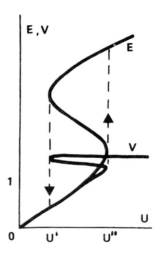

FIG. 4.18

large wind-up. That is why we will focus the analysis on the case $E'_a < 1$ to be called *substantial jump resonance*.

After the substantial jump,

$$E'_a = \frac{1}{|F_0|} U' = \frac{1}{|F_0|} \inf_{E>1} U(E) = \frac{1}{|F_0|} \inf_{E>1} \left| \frac{ET(T+1)}{T} \right|$$

$$= \left| \frac{T_0}{F_0} \right| \inf_{E>1} \left| \frac{EHF}{T} \right| = |M_0| \inf_{E>1} \left| \frac{EH}{M} \right| \qquad (4.15)$$

or, identically,

$$E'_a = |M_0| \left| \frac{E'_b H(E'_b)}{M(E'_b)} \right| \qquad (4.16)$$

Because $E = U/F = UT/M$, the jump in the phase associated with the downward jump in the amplitude can be calculated as

$$\arg E'_a - \arg E'_b = \arg \frac{F'_b}{F'_a} = \arg M \frac{H(E'_b)}{M(E'_b)}$$

$$= \arg M_0 + \arg H(E'_b) - \arg M(E'_b) \qquad (4.17)$$

After substituting the approximate expression for the DF of dynamic saturation (3.28) into (4.15), we obtain

$$E'_a = |M_0| \inf_{E>1} \frac{1.27 - 0.27 E^{-3}}{|M(E)|} \qquad (4.18)$$

or

$$E'_a = |M_0| \frac{1.27 - 0.27 E'^{-3}_b}{|M(E'_b)|} \qquad (4.19)$$

Consider a realistic case of large feedback when $0.5|T_0| \gg 1$ and $|M_0| \cong 1$. In this case $|E'_b| > 2$, because if it is not, from (3.28), $|H| \geq 0.5$ and therefore $|T| \geq 0.5|T_0| \gg 1$ and $|M(E'_b)| \cong 1$; (4.19) becomes incompatible with the conditions for substantial jump $E'_a < 1$. Hence, with reasonable accuracy, the second term in the numerator in (4.18) can be neglected, and we arrive at a very simple formula:

$$E'_a = \frac{1.27}{\sup_{E>1} M} \qquad (4.20)$$

whose denominator can readily be found by plotting the AAPC on the Nichols chart. This is illustrated in Fig. 4.19: the curve $|M| = 1.5$ is tangent to the AAPC, hence $E'_a = 0.85$.

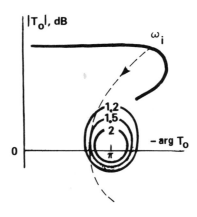

FIG. 4.19

When AAPC trends close to the rectangle of stability margins determined by $x = 10\,\text{dB}$ and $y = 1/6$, the point $(150°, 10\,\text{dB})$ gives $|M| = 2$, which is the largest $|M|$ on this AAPC, resulting in $E'_a = 0.65$.

4.3.3 System with a Nondynamic Nonlinear Link

Because a nondynamic nonlinear link yields $\arg H = 0$, from (4.13),

$$|F^2| = 1 + 2|T_0|\,H\cos\varphi + |T_0^2|\,H^2 \tag{4.21}$$

where $\varphi = \arg T_0$. Further, from (4.12),

$$1 + 2|T_0|\,H\cos\varphi + |T_0|\,E\cos\varphi\,\frac{dH}{dE} + |T_0^2|\,H^2 + |T_0^2|\,EH\,\frac{dH}{dE} = 0 \tag{4.22}$$

so that

$$-\cos\varphi = \frac{|T_0^2|\,H^2 + |T_0^2|\,EH\,dH/dE + 1}{|T_0|\,(2H + E\,dH/dE)} \tag{4.23}$$

for $E = E'_b$ and $E = E''_b$.

Now, let us determine the values of $T_{0\,\text{lim}}$, which correspond to the limit case $E'_b = E''_b = E_b$. Certainly, E_b is the root of an equation obtained by derivation of (4.22):

$$3(\cos\varphi - |T_0|H)\frac{dH}{dE} - |T_0|E\left(\frac{dH}{dE}\right)^2 + (E\cos\varphi - |T_0|H)\frac{d^2H}{dE^2} = 0$$

(4.24)

which follows in the equation:

$$|T_{0\,\text{lim}}| = \sqrt{\frac{3dH/dE + Ed^2H/dE^2}{H^2(3dH/dE + Ed^2H/dE^2) + E(dH/dE)^2(2H + EdH/dE)}}$$

$: E = E_b$

(4.25)

4.3.4 System with Nondynamic Saturation

Substituting expression (3.25) into (4.21), we obtain the equation:

$$U^2 = E^2 + 2.54|T_0|E\cos\varphi + 1.61|T_0|^2 - 0.54|T_0|E^{-2}\cos\varphi$$
$$- 0.686|T_0|E^{-3} + 0.073|T_0|^2 E^{-6}$$

(4.26)

The bifurcation values E_b' and E_b'' are the roots of the equation obtained by combining (3.25) and (4.22):

$$E^8 + 1.27|T_0|E^7\cos\varphi + 0.54|T_0|E^4\cos\varphi + 1.028|T_0|^2 E^3 - 0.218|T_0|^2 = 0$$

(4.27)

from which

$$-\cos\varphi = \frac{E^8 + 1.028|T_0^2|E^3 - 0.218|T_0^2|}{1.27|T_0|E^7 + 0.54|T_0|E^4} \quad : E = E_b' \text{ or } E = E_b''$$

(4.28)

By permutation of (3.25) into (4.25), the limit case value for $|T_0|$ is determined as

$$|T_{0\,\text{lim}}| = 0.494\,E^4 \sqrt{\frac{E_b^3 + 1.7}{E_b^3(E_b^3 - 0.265) - 0.09}} \quad : E = E_b$$

(4.29)

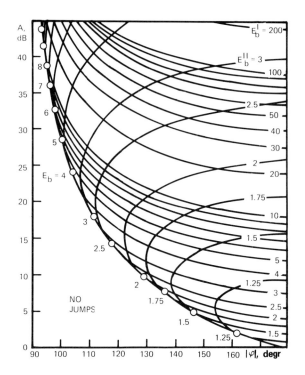

FIG. 4.20

The family of curves corresponding to (4.28) is plotted in Fig. 4.20. Using these curves and (4.26), the family of input bifurcation amplitudes U' and U'' is plotted onto the Nichols chart in Fig. 4.21. The family of curves $E'_a = U'/|F_0| = \text{const} < 1$, which is valid for the substantial jump-resonance analysis, is shown in Fig. 4.22. Using these diagrams, we are able to solve the direct and inverse problems: to find the minimum allowable phase stability margin if the maximum admissible values of jumps are given; and to calculate the phase stability margin utilizing $|T_0|$ and E'_a obtained experimentally. The latter application also appears to be very convenient in practice.

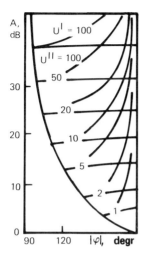
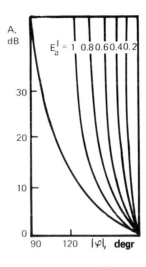

FIG. 4.21 FIG. 4.22

4.3.5 Substantiation of DF Technique for the Jump-Resonance Analysis

We will now prove the consistency of the hypothesis that when $U = U'$, or $U = U''$, the shape of $e(t)$ is nearly sinusoidal.

Provided that this is correct $v(t)$ approximates a trapezoidal shape, thus containing only odd harmonics and having the ratio in amplitudes of the nth harmonic to the fundamental less than $1/n$ (the value $1/n$ relates to the meander).

As shown, for the jumps to exist, $|\arg T_0| > \pi/2$. The average slope of the Bode diagram is then steeper than $-6\,\mathrm{dB}/\mathrm{octave}$, i.e., $|T_0|$ decreases *versus* frequency more rapidly than ω^{-1}. Hence, the ratio in amplitudes of the fundamental to the nth harmonic in the signal fed back to the summing point is at least n^2. In the most important case of $|\arg T_0| > 3\pi/4$, when the jump-resonance is substantial, the value of $|T_0|$ decreases with frequency even faster, as $\omega^{-1.5}$, and the fundamental in the signal fed back is more than $n^{2.5}$ times larger than the amplitude of the nth harmonic.

Because the input is assumed to be sinusoidal, the harmonics in $e(t)$ are those of the return (fed back) signal. Finally, in order to complete the proof, it must be shown that E is of the same order of magnitude as the fundamental of the return signal $ET(E)$. This can be directly seen from Fig. 4.2, with higher accuracy for the events of practical interest related to $U = U'$.

4.3.6 System with Dead Zone Element

If the amplitudes of the signals applied to the inputs of the systems of Fig. 4.12(a,c) are equal, the amplitudes of the signals at the inputs of the nonlinear links are equal as well. Therefore, the already calculated boundary curves for the system with saturation can be mapped onto the T_0'-plane using (4.10), to form the boundary curves for the system with a dead zone link. The calculations show that the boundaries of jump-resonance existence map onto one another and, therefore, if jump-resonance takes place in a system with a saturation link, it will exist after replacing the saturation link by the dead zone, and *vice versa*.

Using the analogy between the feedback system and the interconnection of two two-poles, we can apply the foregoing analysis to the circuit with a nonlinear inductor depicted in Fig. 4.23(a) [63], expressing the return ratio as

$$T_0 = \frac{Z_2}{Z_1} = \frac{R/(1+j\omega RC)}{j\omega L}$$

The slope of the Bode diagram for this T_0 is fairly large, so the filter conjecture is applicable, and therefore $-\pi < \arg T_0 < -\pi/2$. Using the flow chart representation for a nonlinear inductor, as in Fig. 4.23(b,c) (similar to that used in Sec. 3.2.1.), we find the equivalent feedback system shown in Fig. 4.23(d), where v represents saturation and v^{-1} represents dead zone. For this circuit, the dependence $U_L(U)$ is single-valued or triple-valued, depending on the arg T_0.

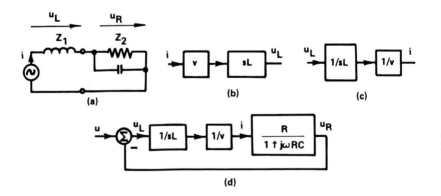

FIG. 4.23

4.3.7 System with Nonlinear Element having Power-Type DF

Consider $H = E^m$, where $m > 0$, (following [12]; the particular case of $m = 2$ was analysed in [174]). Substituting H into (4.21) and multiplying by E gives

$$U^2 = E^2 + 2|T_0| E^{m+2} \cos \varphi + |T_0| E^{2m+2} \tag{4.30}$$

By equating its first derivative to zero,

$$E^{2m} (m + 1) |T_0|^2 - E^m (m + 2) |T_0| \cos \varphi + 1 = 0$$

we obtain an equation for finding E_b' and E_b'':

$$E_b', E_b'' = \left\{ \frac{[(m + 2)/2] \cos \varphi \pm \sqrt{[(m + 2)/2] \cos^2 \varphi - (m + 1)}}{|T| (m + 1)} \right\}^{1/m}$$

As E_b' and E_b'' are real and positive, the radicand must be positive. It equals zero in the limit case of the jump-resonance existing. Hence, the jump-resonance is absent if and only if

$$|\arg T_0| \leq \arcsin [m/(m + 2)] \tag{4.31}$$

which is illustrated by curve 1 in Fig. 4.24.

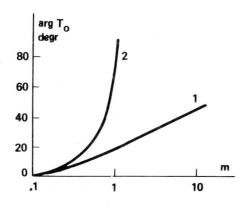

FIG. 4.24

Next, let $H = E^m$, $0 < m < 1$ (the following analysis is not valid for larger m, which necessarily imply a discontinuity in the characteristic $v(e)$). In this case,

$$U^2 = E^2 - 2|T_0| E^{2-m} \cos \varphi - |T_0| E^{2-2m}$$

and, in the same manner, we find

$$E_b', E_b'' = \left\{ |T_0| \frac{2-m}{2} \cos \varphi \pm |T_0| \left[\frac{(2-m)^2}{4} - \cos^2 \varphi - (1-m) \right]^{1/2} \right\}^{1/m}$$

and the condition for the absence of jump-resonance:

$$|\arg T_0| \leq \arcsin [m/(2-m)] \tag{4.32}$$

as plotted in Fig. 4.24 by curve 2. On the Nichols chart, the borders for the area of jump-resonance existence take the form of parallel lines.

We see that under the imposed requirement of eliminating the jump-resonance, any deviation of m from 1 reduces the allowable $|\arg T_0|$, thus decreasing the maximum attainable feedback. This is one of the reasons nondynamic nonlinear correctors are often installed at the input to the nonlinear link of the actuator in order to linearize the total transfer characteristic.

4.3.8 Examples and Exercises

1. In a system with nondynamic saturation, at a certain frequency ω_i, the value of E after the jump is $E_a' = 0.5$, and $|T_0(j\omega_i)| = 50$. Determine $\arg T_0(j\omega_i)$ (use Fig. 4.22).

2. In a system with nondynamic saturation, $|T_0| = 50$ and $\arg T_0 = -5\pi/6$. Determine E_b', E_b'', U', U'', E_a' (use Figs. 4.20, 4.21, and 4.22) and the value of the jump in phase that is associated with the downward jump in amplitude.

3. In a system with saturation, $T_0(j\omega)$ represents Bode optimal cut-off, and at $\omega = 1$, $|T_0| = 50$. Find the amplitude of the third harmonic at the input to the nonlinear link in the state facing the downward jump, across the normalized frequency band $1 < \omega < 4$. What maximum shift in the position of the boundary curves in Figs. 4.20, 4.21, and 4.22 can this harmonic contribute?

4. A small hysteresis exists in a feedback system with saturation. What changes will it cause, qualitatively and quantitatively, in the position of the curves in Figs. 4.20, 4.21, and 4.22?

5. In a system with dynamic saturation, AAPC T on the Nichols chart is tangent to the curve $|M| = 3$. What is the E'_a?

6. Contrary to the system with saturation or dead zone element, in a system with the power-type nonlinear link defined by (4.31) and (4.32), why does the boundary curve of the jump-resonance not depend on $|T_0|$?

7. When the magnetic core of the nonlinear inductor of the circuit shown in Fig. 4.23(a) is fully saturated by means of an auxiliary winding fed by a source of direct current, the characteristics $U_L(U)$ become quintuple-valued [63], as those shown in Fig. 4.2(c). Give an explanation with reasoning similar to that employed in Sec. 4.1.

8. The symmetrical nonlinear link $v(e)$ is described by

$$v = \begin{cases} 100e : e \leq 0.006 \\ 0.6 : 0.006 < e < 0.6 \\ e : 0.6 < e < 1 \\ 1 : e > 1 \end{cases}$$

In the system $|T_0(j\omega)| = 100$, arg $T_0(j\omega) = -160°$. Find the dependence $U(E)$. Find the envelope of $e(t)$ and $v(t)$ when $u(t) = a(\sin \omega_1 t) \cdot \sin \omega t$, $\omega_1 \ll \omega$. Show that application of such a device improves the shape of radio pulses in the same manner as the Schmidt trigger does with video pulses.

4.4 ODD SUBHARMONICS

We will examine the mechanism of generating nth-order odd subharmonics in a lowpass AGS system including a memoryless saturation link with unity threshold.

As will be shown below, the subharmonics in such a system may only become excited if the phase stability margin at the frequency of the subharmonic ω/n is rather small. This implies a steep cut-off of the Bode diagram for T_0. Therefore, the signal at the input to the nonlinear link consists mainly of the lower frequency components: $E \sin \omega t$ and $E_{sb} \sin \omega t/n$.

For the sake of approximate analysis, examine whether the subharmonic oscillation is possible with either small or large amplitudes of E and E_{sb}.

Evidently, with both E and E_{sb} being so small that $E + E_{sb} < 1$, the subharmonics cannot be observed, since the system behaves linearly.

If $E > 1$ and $E_{sb} \ll 1$, the nonlinear link is saturated by the signal and, for small increments given to its input, the link can be seen as an equivalent, linear, time-varying one, i.e., a pulse element with the sampling frequency

ω. Considering the subharmonic as such an increment, the increment at the output of this link is the product of $E_{sb} \sin \omega t/n$ and the Fourier expansion with the fundamental frequency ω, characterizing the pulse element. In this product, the component with the frequency ω/n can only be generated due to the constant component of the Fourier series. This component is certainly real. It follows that the nonlinear link does not create any phase shift for the subharmonic. Therefore, the small-amplitude odd subharmonics cannot occur.

From here, a *corollary* follows: when the parameters of the system or of the input signal are manipulated slowly and continuously, while maintaining the system to be AGS with finite stability margins, the odd subharmonics may only originate by jump, with the steady-state amplitude E_{sb} finite.

Next, if $E_{sb} \gg 1$ (which implies $|T_0(j\omega/n)| \gg 1$), and $E \ll E_{sb}$, the output of the nonlinear link accepts the shape of trapezoidal pulses shown in Fig. 4.25. These pulses are shifted by ψ due to the signal $E \sin \omega t$. If ψ exceeds the phase stability margin at the frequency of the subharmonic and the loop gain at this frequency is more than 1, a steady-state subharmonic oscillation may be excited by creating the appropriate initial conditions.

It can be seen from Fig. 4.25 that $\psi < \pi/2n$. In practical control systems, where the phase stability margin is always greater than $\pi/6$, the subharmonics cannot be observed. In other words, the requirement to eliminate the odd subharmonics does not impose an additional constraint on the feedback system design.

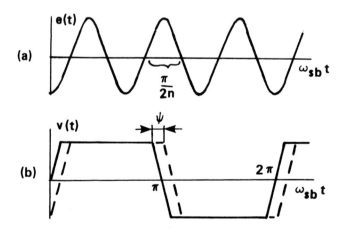

FIG. 4.25

The boundaries for third-order [173, 174] and higher order subharmonics [107] are shown in Fig. 4.26. We can see by comparison of these curves with the plots of Fig. 4.22 that the third subharmonic may not exist if for the jump-resonance at the frequency of the supposed subharmonic $E_a' < 0.5$ (which is required to eliminate wind-up in practical systems).

This rule of thumb, that the requirement to bind E_a' from below automatically excludes the odd subharmonics, appears to remain valid for systems with multiple nonlinear links, although we are not yet able to prove it theoretically.

It may be suspected by extrapolation of the plots of Fig. 4.26 that the second subharmonic might present a real danger in practice. Fortunately, it does not, as will be shown in the next section.

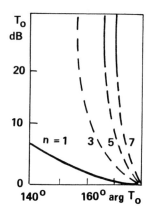

FIG. 4.26

4.5 SECOND SUBHARMONIC

Experiments show that the second subharmonic only occurs in the feedback system if the loop nonlinear link characteristic is asymmetric. Unlike the odd subharmonics, the second subharmonic self-excitation is soft. The amplitude of the subharmonic does not depend on the initial conditions and increases steadily from zero when gradually changing U or ω along any trajectory entering the range in which the subharmonic can be observed.

Since the goal of control system engineering is not to create but to suppress the subharmonics, and because the study of subharmonics is rather complicated, we only determine the boundary of the subharmonic range. The infinitesimal amplitude of the subharmonic at the boundary justifies substitution, for the sake of analysis, of the saturation link by an equivalent time-variable link. Its transfer function equals zero when the nonlinear link is saturated, and one otherwise.

Due to filter-like responses of the loop linear links, two signal components dominate at the input to the nonlinear link, the subharmonic $E_{sb} \cos \omega_{sb} t$ and the signal $e(t) = E \cos(2\omega_{sb} t - 2\gamma)$. Assuming the threshold for the positive polarity of the signal is 1, and for the negative polarity is $e_{s2} < 1$, the nonlinear link becomes saturated over the intervals of the clipping angles shown in Fig. 4.27(a):

$$\chi_1 = \arccos E^{-1} \tag{4.33}$$

and

$$\chi_2 = \arccos e_{s2} E^{-1} < \chi_1 \tag{4.34}$$

centered alternately at $\omega_{sb} t = \gamma, \gamma + \pi/2, \gamma + \pi$, *et cetera*. The time-dependent transmission coefficient of the equivalent linear link is shown in Fig. 4.27(b).

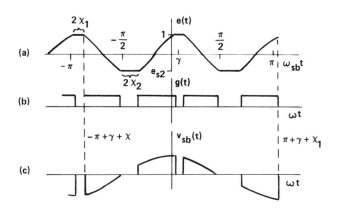

FIG. 4.27

The increment $E_{sb} \cos \omega_{sb} t$ at the input to this link produces the increment at the link's output shown in Fig. 4.27(c). The first term coefficients of its Fourier expansion are given by

Forced Oscillation

$$\text{Re } \mathbf{V}_{sb} = \frac{E_{sb}}{\pi} \int_{-\pi+\gamma-\chi_1}^{\pi+\gamma-\chi_1} \frac{dv}{de} \cos^2 \omega_{sb}t \, d\omega_{sb}t = E_{sb} - \frac{E_{sb}}{\pi} \Bigg[\int_{-\gamma-\chi_1}^{-\gamma+\chi_1} \cos^2 \omega_{sb}t$$

$$\times d\omega_{sb}t + \int_{\gamma-\pi-\chi_1}^{\gamma-\pi+\chi_1} \cos^2 \omega_{sb}t \, d\omega_{sb}t + \int_{\gamma-\frac{\pi}{2}-\chi_2}^{\gamma-\frac{\pi}{2}+\chi_2} \cos^2 \omega_{sb}t \, d\omega_{sb}t$$

$$+ \int_{\gamma+\frac{\pi}{2}-\chi_2}^{\gamma+\frac{\pi}{2}+\chi_2} \cos^2 \omega_{sb}t \, d\omega_{sb}t \Bigg] = E_{sb} \Bigg[1 - \frac{2\chi_1 + 2\chi_2}{\pi} + \frac{\cos 2\gamma}{\pi}$$

$$\times (\sin 2\chi_1 - \sin 2\chi_2) \Bigg] \tag{4.35}$$

$$\text{Im } \mathbf{V}_{sb} = \frac{E_{sb}}{\pi} \Bigg[\int_{-\gamma-\chi_1}^{-\gamma+\chi_1} \cos \omega_{sb}t \sin \omega_{sb}t \, d\omega_{sb}t + \int_{\gamma-\pi\chi_1}^{\gamma-\pi+\chi_1} \cos \omega_{sb}t \sin \omega_{sb}t \, d\omega_{sb}t$$

$$+ \int_{\gamma+\frac{\pi}{2}-\chi_2}^{\gamma+\frac{\pi}{2}+\chi_2} \cos \omega_{sb}t \sin + \omega_{sb}t \, d\omega_{sb}t + \int_{\gamma-\frac{\pi}{2}-\chi_2}^{\gamma-\frac{\pi}{2}+\chi_2} \cos \omega_{sb}t \sin \omega_{sb}t \, d\omega_{sb}t \Bigg]$$

$$= E_{sb} \frac{\sin 2\gamma}{\pi} (\sin 2\chi_2 - \sin 2\chi_1) \tag{4.36}$$

The modulus of the equivalent transfer function of the nonlinear link for the second subharmonic now is

$$|H| = [\text{Re}^2 \mathbf{V}_{sb} + \text{Im } \mathbf{V}_{sb}]^{1/2} / E_{sb} \tag{4.37}$$

and its phase is

$$\psi = \arctan \frac{\text{Im } \mathbf{V}_{sb}}{\text{Re } \mathbf{V}_{sb}} \tag{4.38}$$

Figure 4.28 displays the dependence of $1/|H|$ on $-\psi$, for various constant E and e_{s2}, the points along the curve differing in γ. The boundary curves permit us to determine whether the subharmonic oscillation is possible with the given e_{s2}, E, and ω_{sb}, by drawing the Nyquist diagram for T_0 on the plane of Fig. 4.28 and examining whether the point $T_0(j\omega_{sb})$ falls within the relevant curve.

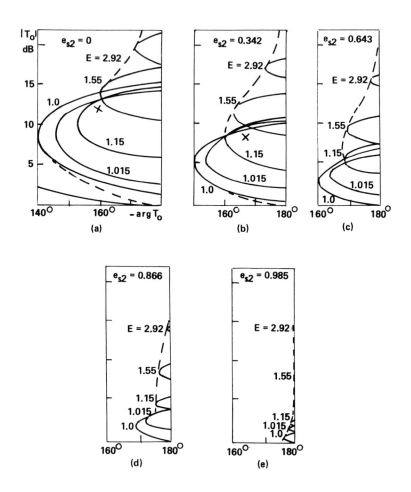

FIG. 4.28

Experiments have confirmed the theoretical calculations [106]. It can be seen that the second subharmonic does not exist in properly designed control systems, where $|\arg T_0(j\omega_{sb})| < 5\pi/6$.

We conclude that in single-loop systems with saturation neither the odd nor the second subharmonics present real danger, since meeting the mandatory requirement for eliminating the wind-up and the substantial jump-resonance automatically excludes the subharmonics. Although the details of the subharmonic self-excitation in systems with multiple or dynamic nonlinear links are not yet well known and may differ from those studied (for example, the second subharmonic may start with a jump [106]), the above conclusion seems to be applicable in almost all practical situations.

Chapter 5
Nonlinear Dynamic Compensator

5.1 LOOP REGULATION

5.1.1 Variable Loop Gain

The gain of the plant might vary extensively. In addition to these random variations, the closed-loop system transmission function K_β is often regulated by changing β. This results in large variations in $|T_0|$ within the working frequency band: over 15 dB, and even 30 dB. At the same time, the noise effect at the actuator input and other limitation on the value of the crossover frequency ω_b typically vary within a smaller range, so the maximum feedback available with proper loop compensator (optimized for the given values of the plant and β-path transmission functions) vary within a smaller range, typically less than 10 dB.

If the loop compensators are time-invariable, they must shape the loop gain response so as to provide stability for the case of maximal gain of the variable links. Hence, the Bode diagram for the case of minimal gain of the variable links is not optimal, and the feedback related to this case is smaller than the maximum available.

This feedback is larger in adaptive systems, with time-variable loop compensators. Generally, since the plant variations differ statistically from the tracked signal, this statistic can be used to adjust (tune) the compensator; in particular, increasing its gain in the working frequency range, either smoothly or by switching. The generic form of the adaptive system is illustrated in Fig. 5.1. Often, the only large components of the loop gain variations are those that are slowly changing in time, and low-speed tuning of a variable compensator placed at the input to the plant is sufficient for significant increase of the feedback.

It is generally required to reduce the size, weight, and cost of the actuator. Therefore, the actuator cannot be excessively powerful, and it is accepted in practice that the actuator becomes overloaded (saturated) from time to time by the signals of greatest amplitudes. The conditions of small and large signals at the actuator input are worth considering separately, since they specify different features of the control laws.

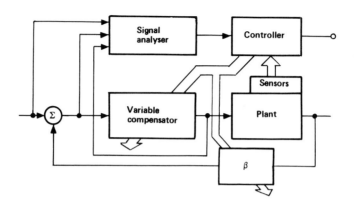

FIG. 5.1

To reduce the small amplitude errors rapidly and accurately, it is desirable to maintain the loop gain maximized over the operational frequency band. When the plant parameters vary, a variable compensator may be employed to optimize the loop gain response for each particular set of values of the plant parameters.

When the input signal changes abruptly by a large value, the error at the input to the actuator exceeds the threshold of saturation. The time to reduce the large error to a level below the saturation threshold depends upon the maximum available output power of the actuator and on the value of the wind-up, but almost not at all on the value of the loop gain. Thus, in the state of large error, the variable compensator of Fig. 5.1 should optimize the loop gain frequency response so as to eliminate the wind-up for any particular shape of the plant response within the tolerances limit, and guarantee the system's global stability and ruggedness.

As an example, consider the phase locked loop (PLL) employed in digital communication systems for extracting the clock from the flow of digital input signal. To minimize the jitter, the feedback in the PLL must be large, 40 to 100 dB, within the relatively narrow bandwidth of the dominant frequency components of the jitter. In such a system, however, the acquisition frequency band is small so that at initialization the input signal clock must be close to the frequency of the voltage-controlled oscillator (VCO). To expand the acquisition band, the shape of the Bode diagram for the PLL should be modified so as to provide moderate feedback, which is spread over a rather wide frequency band. The trade-off between reduction of the jitter and the width of the acquisition band can be resolved in an adaptive system by separately optimizing the shape of the loop gain response for these two modes of operation: first, when the error is large and the VCO is out of synchronism with the data clock; and, second, when the output is synchronized with the input. The transition from the first mode to the second ought to be smooth so as not to lose synchronism, and global stability must be provided to prevent a random large-amplitude error from causing the system to diverge from synchronism.

Flight control is another area of adaptive control application. The dynamics of an aircraft or a missile are greatly influenced by the amount of fuel in tanks, the yaw angle, and the atmospheric density related to the flight altitude. These and some other variations are much slower than the flight regulation, allowing for an auxiliary low-speed closed loop of adaptation that continuously optimizes the parameters of the main control loop. If the process of adaptation needs to be rapid, feedback maximization in the adaptation loop becomes a limiting factor, and the system becomes rather complicated.

As an alternative or addition to the linear variable compensator, a simpler and faster nonlinear dynamic compensator (NDC) may be employed to adjust the forward-path transmission function while "sensing" the value, spectrum, *et cetera* of the error. The study of NDC will start in Sec. 5.1.4.

5.1.2 Switching in the Compensator

The linear link transmission can be changed by a switch [76, 77], as an alternative to smooth regulation. Such regulators are employed in radar tracking systems, in the PLL of communication and power systems, and in many other industrial control systems to reshape the loop gain frequency response during the tracking process. In this section, however, we will only discuss the application of a switching regulator for ensuring global stability in a Nyquist-stable system. The experimental circuits first described by Oizumi and Kimura [151] are diagrammed in Figs. 5.2 and 5.3, and the Nyquist diagrams in Figs. 5.4 and 5.5, respectively. The unmarked blocks represent some linear circuits.

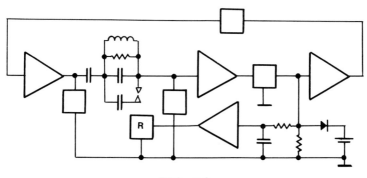

FIG. 5.2

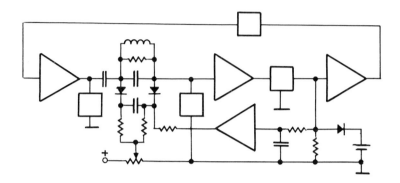

FIG. 5.3

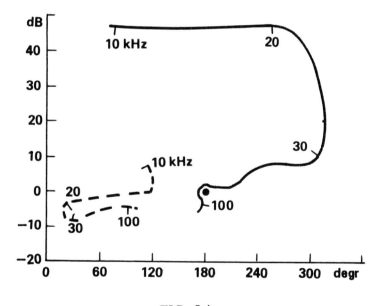

FIG. 5.4

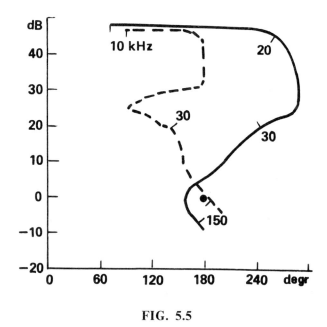

FIG. 5.5

As long as the signal amplitude at the input to the ultimate stage is below the saturation threshold, the feedback within the operational band is large, as can be seen from the locus of the return ratio T_0 shown by the solid lines. Signals of larger amplitudes are detected, averaged in RC filters, and after amplification applied to the relay R (or to a diode switch), changing the configuration and the frequency response of the interstage circuit. With the switch on, the return ratio is plotted on the diagrams by dashed lines.

In the system of Fig. 5.4, no self-oscillation can take place. This keeps the contacts of the relay either permanently closed or permanently open, since in the first case the system is AS, and, in the second case, should the system self-oscillate, a signal that is growing exponentially in amplitude would eventually turn the relay on. Thus, the only conceivable type of self-oscillation is that which makes the contacts of the relay vibrate. Self-oscillation may be eliminated by making the relay hold-up time long enough to allow the signal in the resulting AS system to attenuate below the threshold of switching.

The long hold-up time, however, is not desirable, since a large sporadic pulse to the system input cuts down the feedback for this duration. The hold-up time is shortened in the system of Fig. 5.6, particularly due to the peculiar shape of the Nyquist diagram in the state of closed contacts. In this state, the system is unstable, but only generates a high-frequency signal of small amplitude, which cannot turn the switch on after the hold-up time expires.

The type of Nyquist-stable, globally-stable system described, which consists of a switching circuit and a circuit for recognizing self-oscillation, possesses two general drawbacks: the time of self-oscillation recognition is large; and, secondly, the switching produces extensive transients. Faster and smoother regulation of the return ratio is possible when using NDC.

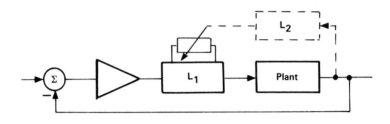

FIG. 5.6

5.1.3 Quasilinear Variable Compensator

As we saw in Chapter 2, the maximum available feedback increases with the steepness of the slope of the Bode diagram and reduction in the phase stability margin. Bode [22] suggested a method for implementation of a system wherein the phase stability margin equals zero. The design, later discussed and developed by several authors [66, 70, 19], was employed, in particular, in the autopilot control loop. Figure 5.6 shows such a system, and Fig. 5.7 illustrates the application of the Nyquist diagram that passes through the critical point (since $y = 0$). The transmission function of quasilinear link L_1 changes under the control of the output signal processed in the nonlinear link L_2. At the frequency at which $|T| = -1$, the return difference possesses a purely imaginary zero, and the system oscillates permanently. The level of this oscillation is kept low by automatically reducing the gain of the link L_1 (by employing linear variable elements such as FET, IC, PIN-diode, or, as in the original Bode version, thermoresistor), and the system is "practically stable."

In the case of the thermoresistor, or thermistor, it is possible for the system to operate without the link L_2 if the network L_1 supplies enough power to heat up the thermistor to change the loop transmission function so that the return ratio for larger power levels does not approach the critical point at any frequency other than at the higher frequencies, where $|T_0|$ is slightly larger than 1. Then, the rather small signal of self-oscillation reduces $|T_0|$ to 1 (by heating the thermistor), thus preventing the oscillation from further increase. This oscillation is almost not observable at the plant's output due to the plant's lowpass type of frequency response. The disadvantages of using a thermistor are its slow response to the changes of the input signal and the dependence of its resistance on the ambient temperature.

Replacement of the thermistor by a nonlinear two-pole forms a NDC.

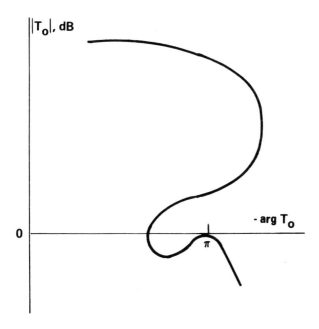

FIG. 5.7

5.1.4 Nonlinear Dynamic Compensator

A nonlinear dynamic compensator (NDC) at the input of the actuator improves the system performance when there are large errors in amplitude. It reduces the wind-up and maximum swing of the jump-resonance without sacrificing the available feedback in the small-error mode of operation. In analog operation, the simplest NDC consists of a capacitor shunted by a diode, or a back-parallel combination of diodes, as depicted in Figs. 5.8 and 5.9. In digital form, the NDC may be operated as a nonlinear transversal filter.

In Fig. 5.8, a simplified circuit diagram of the output stage (actuator) of a feedback amplifier is shown with NDC consisting of two prebiased diodes (one of them, the diode grid-cathode) shunting the stray interstage capacitance. This circuit was employed in the repeaters of the telecommunication system "J" developed by Bell Laboratories in the 1930s. For signals whose amplitude exceeds the saturation threshold of the last tube, the diodes open and the phase lag of the describing transmission function of the interstage circuit decreases by ~30°. This is enough to eliminate substantial jump resonance, which can be evaluated with the methods given in Chapter 4.

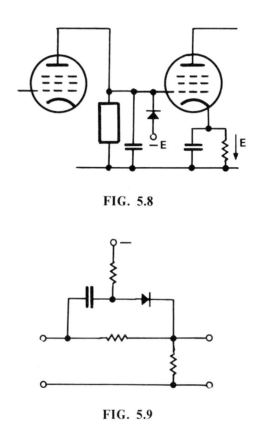

FIG. 5.8

FIG. 5.9

Figure 5.9 shows the NDC employed in temperature controllers for industrial furnaces, eliminating the wind-up by introducing a parallel path for large-level signals through the capacitor, thus introducing a phase lead of 20° to 60° in the describing function of the loop transmission.

Different types of NDCs employ multipliers, varying reactive elements, and links with active hysteresis [37, 48, 162, 164, 167, 171]. No comprehensive comparative study of these NDCs exists. In the author's opinion, most of these links are inferior to the NDC that combines one or two nonlinear nondynamic elements with linear dynamic elements in the manner illustrated in Fig. 5.8, but with a higher-order linear part, or an equivalent nonlinear transversal filter. Such NDCs provide large phase lead and may be applied to improve large-level signal dynamics and to ensure global stability in Nyquist-stable systems.

5.2 DESCRIBING FUNCTION APPROACH TO NDC DESIGN

5.2.1 Generalities

A Nyquist-stable feedback system will be considered. It contains two nonlinear links in the loop: the NDC and the ultimate power stage of the forward path (actuator). The latter presumably possesses the characteristic of nondynamic saturation. As shown in Fig. 5.10, these two links with the corresponding DF, H_1 and H_2, are separated along the loop by the two linear links M_1 and M_2. It is assumed that if self-oscillation takes place in this system, the signals at the inputs to the nonlinear links are nearly sinusoidal with amplitudes E_1 and E_2, due to the lowpass properties of the linear links M_1 and M_2. These amplitudes are normalized relative to the threshold of saturation in the link H_2 (this threshold, therefore, is viewed as 1).

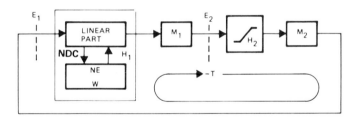

FIG. 5.10

The NDC consists of a linear part and a nonlinear nondynamic element NE, either a two-port or a two-pole. The NDC is required to perform as a linear link with the transmission function $H_1(0)$ for sinusoidal signals with frequencies $\omega \leq 1$ and amplitudes $E_1 < E_{1s}$. For these signals, the threshold is required to be

$$E_{1s} \geq |H_1(0) M_1|^{-1} \tag{5.1}$$

to render the maximum undistorted signal amplitude $E_{1s}|H_1(0) M_1|$ at the input to H_1 exceeding the threshold of 1. Introducing such a NDC in the loop preserves the maximum available signal at the system's output.

In the following discussion, emphasis will be placed on understanding the basic considerations. We presume that when a crude initial solution is attained with simplified approximate relations, it can be further improved by using a local optimization technique.

5.2.2 Stability Conditions

Through use of the DF technique, stability conditions for the system of Fig. 5.10 can be studied by making a cross section at the input to either the NDC or the link H_2, and applying a sinusoidal signal with amplitude E_1 (or, respectively, E_2) to the input of the broken loop, thus generating either of the DF expressions for the return ratio T, (5.2) or (5.3):

$$-H_1(E_1, j\omega) \cdot M_1(j\omega) \cdot H_2[E_1 H_1(E_1, j\omega) \cdot M_1(j\omega)] M_2(j\omega) \qquad (5.2)$$

$$-H_2(E_2) \cdot M_2(j\omega) \cdot H_1[E_2 H_2(E_2) \cdot M_2(j\omega), j\omega] M_1(j\omega) \qquad (5.3)$$

If no sets of values of ω and E_1 (or E_2) exist that result in the equality $T = -1$, no periodic self-oscillation can occur. It is known from practical experience that an aperiodic self-oscillation also should not be expected. The system is then considered AGS.

Expression (5.3) can be displayed with a family of curves on the Nichols chart. Particularly, amplitude-phase characteristics (APC$_1$) or amplitude-amplitude-phase characteristics (AAPC$_1$) can be of use, as Fig. 5.11 illustrates. Each curve of APC corresponds to some fixed value of E_1, and each curve of AAPC$_1$ corresponds to a certain specified frequency. The thicker line represents $T_0 = M_1 M_2 H_1(0)$. Increasing E_1 produces a positive increment of the phase shift of the NDC that causes all the AAPC curves to avoid the rectangle of amplitude and phase stability margins defined by x, x_1, and $y\pi$.

If the loop is broken at the input to H_2, the similarly defined APC$_2$ and AAPC$_2$ should be used. It is important to realize, however, that AAPC$_1$ and AAPC$_2$ referred to the same frequency do not necessarily coincide, except at the two points indicated in Fig. 5.12: in the linear state of operation, i.e., on T_0, and at the 0 dB level. The first is trivial. The second follows from the fact that H_1 and H_2 depend on the amplitude and not the phase of the signal applied to the link's input. Thus, when the loop is closed and the state $|T| = 1$ is created — we will term this state *critical* and apply the subscript c to the related values (e.g., $T_c = T : |T| = 1$) — then, the signal with the amplitude $E_1 = E_{1c}$ applied at the input to H_1 induces the signal with the amplitude $E_2 = E_{2c}$ to appear at the input of H_2, and *vice versa*.

Therefore, although either of the expressions, (5.2) or (5.3), is applicable for the purposes of analysis, the amplitude stability margins of common definition are not invariant to the location of the loop cross section.

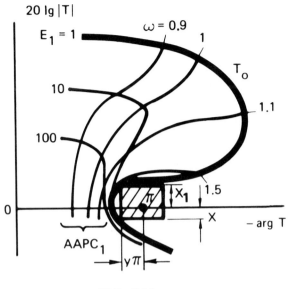

FIG. 5.11

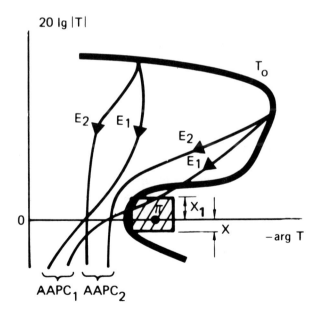

FIG. 5.12

5.2.3 Phase Stability Margin

Below, we will only consider the systems with characteristics similar to those shown in Figs. 5.11 and 5.12, and particularly those with $|T_0|$ decreasing monotonically *versus* ω. The locus of such a $T_0(j\omega)$ on the Nichols chart makes a single intersection with the 0 dB line. This point is located to the left from the point $[(1-y)\pi, 0\,\text{dB}]$. The robustness requirement for the system with certain loop gain tolerances are formulated as:

$$|\arg T_c| < (1-y)\pi \tag{5.4}$$

at all frequencies and all admissible values of $|M_1|$ and $|M_2|$. Such a requirement is irrelevant to the location of the loop cross section, since, as noticed previously, AAPC$_1$ and AAPC$_2$ intersect at the critical state.

Suppose for the sake of evaluating $\arg T_c$ at a certain frequency ω_i that all the nonlinear nondynamic elements (*NE*) of the system, each having the respective DF w_1, w_2, w_3, *et cetera*, are replaced by linear elements equal to the values that these *NE* DF accept under the critical state condition existing at ω_i. We denote these values as $w_{1c}(\omega_i)$, $w_{2c}(\omega_i)$, *et cetera*. The obtained linear system yields the same $T_c(j\omega)$ as the original nonlinear one. Therefore, using the Bode segment method is allowed for the calculation of $\arg T_c(\omega_i)$ from the gain-frequency diagram plotted for this equivalent linear circuit. In contrast to the linear system analysis, a set rather than a single Bode diagram must be created to calculate the frequency response of $\arg T_c$.

Most of the NDCs studied in this text contain a single *NE*. Over the entire frequency range where $|T_0| > 3$, the amplitude of the signal applied to this *NE* is typically large, and w_c is fairly constant. This simplifies the calculation of the NDC's phase shift. The values of w_c are frequently small by comparison with the associated linear elements. For example, if the diodes constituting a two-pole dead zone *NE* are open, their DF conductance is much larger than the driving-point impedance faced by the *NE*. Then, assigning the value of 0 to w_c does not cause a noticeable error. This being so, the same equivalent circuit, and therefore the same gain-frequency response curve are applicable for calculating $\arg T_c$ in a rather large frequency band.

5.2.4 Design Constraints

As indicated in Fig. 5.11, $|\arg T_0| < (1-y)\pi$ when $0.3 < |T_0| < 2$. This takes place in the neighborhood of ω_b, i.e., at the frequencies $\{\omega : 20\log|T_0| \in [-x, x_1]\}$. At these frequencies, (5.4) is satisfied, even when the NDC is replaced by a linear link having a transmission function $H_1(0)$.

Nonlinear Dynamic Compensator

The nonlinearity of the NDC is essential to satisfy (5.4) in the frequency range where $20 \log |T_0| \gg x_1$ and $|\arg T_0| > (1-y)\pi$. In this region, which we call the range of importance, the increment of the phase of H_1:

$$\psi_c = \arg H_{1c} - \arg H_1(0) \tag{5.5}$$

needs to be large.

The DF $H_1(w)$ can be represented as*

$$H_1(w) = \alpha(w+\beta)/(w+\gamma) \tag{5.6}$$

with the parameters α, β, and γ relating to the linear part of the NDC. Figure 5.13 displays the flow chart corresponding to (5.6). The two types of NDC, with either the two-port or the two-pole *NE*, are shown respectively in Figs. 5.14 and 5.15.

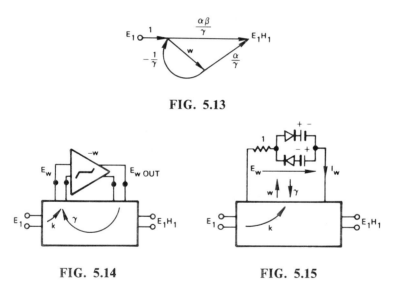

FIG. 5.13

FIG. 5.14 **FIG. 5.15**

The dependencies $w(E_w)$ and $w(E_{w\,out})$ are given by the *NE* characteristic, where E_w is the amplitude of the signal at the input to the *NE*, and $E_{w\,out}$ is the signal amplitude at its output. Therefore, to obtain the function $H_1(E_1)$

---

*We neglect here the error caused by higher harmonics; in ψ_c, they typically contribute less than 10°.

we need to find the dependency of either E_w or $E_{w\,out}$ on E_1. For the sake of simplicity, we consider the cases of the four-pole and two-pole NE separately.

Consider first the system of Fig. 5.14. Let us set $w = -\gamma$, then from (5.6) $H_1 = \infty$. Under this proviso, the main determinant is 0 and the return ratio T_n in the local feedback of the NDC must equal -1, implying the transmission coefficient of the feedback path β_n to be equal to $-1/\gamma$. Therefore, generally, $T_n = w/\gamma$. The output of the NE is, then,

$$E_{w\,out} = k\gamma T_n / (T_n + 1) \tag{5.7}$$

where k stands for the voltage transfer coefficient of the linear part in the direction shown in Fig. 5.14. The DF of the NE is, therefore,

$$w = E_{w\,out}/E_w = k\gamma(E_1/E_w) T_n / (T_n + 1) \tag{5.8}$$

Consider next the system of Fig. 5.15. Setting $w = -\gamma$ gives $H_1 = \infty$, which implies zero main determinant and, therefore, zero contour impedance. From the latter condition, the driving-point admittance seen from w would be γ. Therefore, the return ratio for the two-pole element w (recall Sec. 1.1.3) is $T_n = w/\gamma$. The current that flows through the NE is

$$I_w = kE_1/(w + \gamma) = k\gamma E_1 T_n / (T_n + 1)$$

where k now accepts the meaning of the voltage transfer coefficient from E_1 to the NE, i.e., the ratio of the voltage appearing at the open poles to which w must be connected to the voltage E_1. For $w = I_w/E_w$, expression (5.8) again takes place.

The following study is limited to the intrinsically globally stable NDC that meets the four constraints listed below, in order to simplify the theory and the design procedure. The results that could be attained under these limitations are sufficiently good for practice. The imposed constraints are these:

a. $H_1(w,s)$ possesses neither poles nor zeros in the right half-plane of the operational variable s, i.e., the NDC is stable by itself and its transfer function is of minimum phase type.

b. In the frequency range of importance:

$$\text{Sign Im } \beta = -\text{Sign Im } \gamma \tag{5.9}$$

c. In NDC, the NE is a dead zone one, so that

$$\{w = 0 : E_1 \leq E_{1s}\} \text{ and } \lim_{E_1 \to \infty} w = 1 \tag{5.10}$$

and, therefore,

$$H_1(0) = \alpha\beta/\gamma \tag{5.11}$$

$$H_1(w) = H_1(0)(1 + w/\beta)/(1 + w/\gamma) \tag{5.12}$$

d. It can be seen from (5.5) and (5.12) that, for ψ_c to be substantial, the ratios w_c/β and w_c/γ need to be large. Therefore, in the frequency range of importance, it should be required:

$$w_c > 3|\beta|; \quad w_c > 3|\gamma| \tag{5.13}$$

In this case, $T_n/(T_n + 1) \cong 1$, and (5.7) and (5.8) simplify.

Next prove this statement: with limitations (a) and (b) satisfied, $\arg H_1$ depends monotonically on w, for all real w.

Proof: The bilinear function of (5.6) conformally maps the complex plane w onto the complex plane H_1. The images of $-\beta$ and $-\gamma$ of the w-plane are respectively 0 and ∞ on the H-plane. Because $-\beta$ and $-\gamma$ are located on different sides of the real axis on the w-plane, then, on the H-plane, one of the two points, 0 and ∞, must be inside and the other outside of the circumference, which is the image of the real axis of the w-plane. Hence, this circumference is of finite radius and encompasses the origin.

When w varies monotonically along the real axis, the image of w moves monotonically along the circumference. From this, we can see that $\arg H_1$ varies monotonically as well, *Q.E.D.*

Therefore, in order to achieve maximal changes in $\arg H_1$, the changes in w must be maximized. This is achievable with the relay NE, but the switching may produce undesirable transients. Under the requirement that the system's transient response be smooth, the dead zone type of NE is appropriate, which meets condition (c). With such a NE, from (5.5), (5.11), and (5.12), we have

$$\psi_c = \arg \frac{(w + \beta)/\beta}{(w + \gamma)/\gamma} = \arg \frac{w + \beta}{w + \gamma} \frac{\gamma}{\beta} \tag{5.14}$$

As proved, ψ_c cannot exceed 2π. Let us find what values of β and γ are to be recommended for practice, and what maximal ψ_c can be practically achieved.

Figure 5.16 illustrates the meanings of the angles $\psi_1 = \arg(w_c + \beta)/\beta$ and $\psi_2 = \arg(w_c + \gamma)/\gamma$. We can see that $\psi_c = \psi_1 + \psi_2$ approaches 2π if only Im β and Im γ are both small compared with w_c. If these imaginary parts are, however, excessively small, the sensitivities of H_1 to variations in β and γ become excessively large at the level of a signal that relates to w being close to $-\mathrm{Re}\,\beta$ or $-\mathrm{Re}\,\gamma$. To prevent this sensitivity from being large, we introduce phase safety margins, let us say, of $\pi/6$ described by

$|\arg \beta| \leq 5\pi/6$

$|\arg \gamma| \leq 5\pi/6$

i.e.,

$w_c > 3|\beta|$

$w_c > 3|\gamma|$ (5.15)

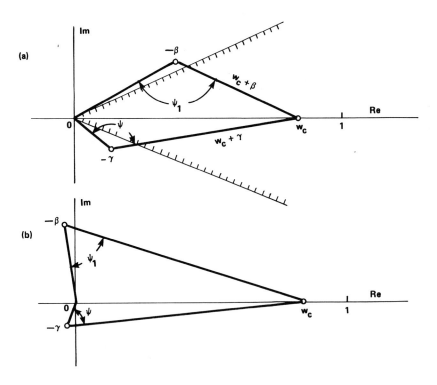

FIG. 5.16

Taking these margins, $\psi_c < 1.7\pi$; but even smaller values of ψ_c, as small as 0.75π, are quite adequate for most applications. Therefore, it is sufficient to use

$$|\arg \beta| \cong \pi/2$$
$$|\arg \gamma| \cong \pi/2 \qquad (5.16)$$

as shown in Fig. 5.16(b).

It will be shown in the next section that, in particular, when $|\arg \beta| = \pi/2$ and $\arg \gamma = -\arg \beta$, the modulus $|H_1|$ monotonically depends on w (as well as ψ_c), thus simplifying the NDC design.

With the condition (5.15) met, $H_{1c}/H_1(0) \cong \beta/\gamma$, and $T_c \cong T_0 H_2 \beta/\gamma$. Since, further, $|T_c| = 1$ and $H_2 = 1$, then

$$|\beta/\gamma| = |T_0| \qquad (5.17)$$

In other words, at the frequencies where $|T_0|$ is large (the frequencies 0.9; 1; 1.1 in Fig. 5.11), $|H_{1c}|$ may be much smaller than $|H_1(0)|$, and AAPC$_1$ may traverse the Nichols chart. At the higher frequencies of the band of importance, the AAPC$_1$ must start to veer left while maintaining $|H_1|$ constant (as at $\omega = 1.5$ in Fig. 5.11). For this reason, it is worth choosing

$$|\beta/\gamma| \cong 1 : 20 \log |T_0| = x_1 \qquad (5.18)$$

As assumed, both $H_1(0, j\omega)$ and $H_1(w, j\omega)$ are stable MP functions. Then, $\beta/\gamma = H_1(0)/H_{1c}$ is MP as well, and in order to satisfy (5.16), the slope of the Bode diagram for β/γ ought to be of $-12\,\text{dB}/\text{octave}$. Therefore, considering also (5.18),

$$|\beta/\gamma| > 1 \qquad (5.19)$$

throughout the frequency range of importance.

Figure 5.17 illustrates for $\beta = \omega'/j\omega\, 10^{1/2}$ and $\gamma = 1/10\beta$ that the conditions listed above are not contradictory. Figure 5.18 represents the loci for $H_1(w)/\alpha$, with w varying from 0 to 1.

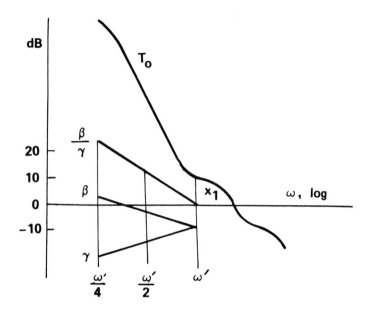

FIG. 5.17

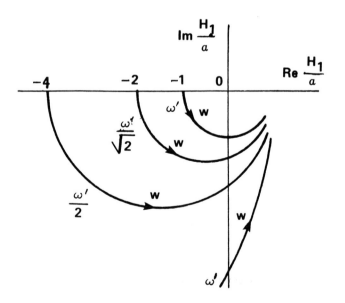

FIG. 5.18

5.2.5 Amplitude Characteristics

In the following simplified analysis we assume that $|\beta/\gamma| \gg 1$. Although this inequality only occurs at lower frequencies, still $|\beta/\gamma| > 1$ throughout the frequency range of importance. Here, the following analysis is applicable, being approximate with considerable accuracy.

The amplitude of the fundamental at the output of the NDC is $E_1 H_1(E_1)$. This function is approximated in a piecewise manner in Fig. 5.19. Let us go through the three sections of this approximation.

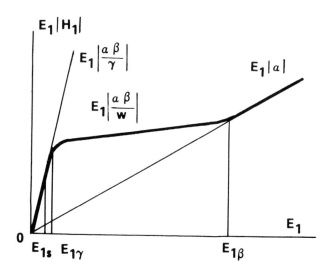

FIG. 5.19

At the first section, small amplitudes $E_1 \leq E_{1s}$ lead to $w = 0$, so from (5.3) $H_1 = \alpha\beta/\gamma$.

At the third (last) section, where E_1 is large, $H_1 \cong \alpha$ (recall that $|\beta| \ll 1$ and $|\gamma| \ll 1$).

At the end of the first section, where E_1 reaches $E_{1\gamma}$ at which $w = |\gamma|$, the output amplitude $|E_1 H_1| = E_1|\alpha\beta/(w+\gamma)| = E_1|\alpha\beta/\gamma| \, 2\sin[(\pi - \arg\gamma)/2]$. When (5.16) is satisfied, the above value differs from the piecewise-linear approximation value of $E_1|\alpha\beta/\gamma|$ by a factor of $\sim\sqrt{2}$, i.e., by only 3 dB.

At the end of the second section, E_1 reaches $E_{1\beta}$ such that $w = |\beta|$. Here, the module of the transmission DF is $|H_1| \cong |\alpha(w+\beta)/\beta| = 2|\alpha|\sin[(\arg\beta)/2]$, which differs from the asymptotic value $|\alpha|$, similarly, by a factor of $\sqrt{2}$.

Finally, within the interval $E_1 \in [E_{1\gamma}, E_{1\beta}]$

$$E_1 H_1 \cong E_1 \alpha\beta/w \tag{5.20}$$

The latter expression involves w, which, in turn, depends on E_1, and therefore it needs further examination.

From (5.12) and (5.13), it follows that the dependence of $|H_1 E_1|$ on E_1 is inverse with respect to the characteristic of the NE (dead zone). Then, $|E_1 H_1|$ monotonically increases over the second interval with a relatively small derivative.

Next, analyze the condition of critical state and determine the amplitude stability margin.

The system of Fig. 5.10 is regarded as composed of two cascaded links, the amplitudes of the input to the first being E_1, and to the second E_2. Figure 5.20 displays these links' transfer characteristics $E_2 = E_1 |M_1 H_1(E_1)|$ and $E_1 = E_2 |H_1(E_2) M_2|$. The upper limit for the amplitude of the fundamental at the output of the link H_2 is $4/\pi \cong 1.27$, and, therefore, at the output of the link M_2, it is $1.27|M_2|$.

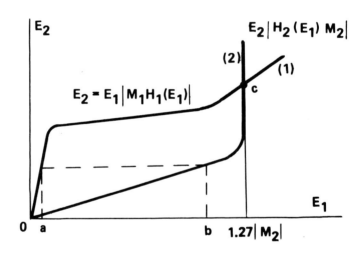

FIG. 5.20

With any given $E_1 = a$, $|T|$ can be readily found as b/a, as shown in Fig. 5.20. The critical state being the solution of the equation $|H_1 M_1| = 1/|H_2 M_2|$ corresponds to the point of crossing of the curves.

At frequencies where $|T_0| \gg 1$, i.e., where $|H_1(0) M_1| \gg 1/|M_2|$, curve 1 is steeper near the origin than curve 2, and shallower at large values of E_1. Hence, the number of intersections of the curves is odd (not including the origin), in fact, one or three. With the constraints of (5.15) met, the curves cross only once.

It was assumed that throughout the frequency band of operation, in the critical state, the signal is limited in the link H_2. Thus, curve 1 crosses curve 2 at its vertical ray. Therefore,

$$E_{1c} = 1.27 |M_2| \tag{5.21}$$

The values of E_{1c} and, consequently, of ψ_c increase with increasing $|M_2|$, although the maximal admissible $|M_2|$ is limited by the above-mentioned condition that within the operational band the saturation should start first in the plant.

At frequencies in the band of importance, off the band of operation, it is advisable that the signal is precluded from being limited in the link H_2 by keeping $E_2 < E_{2s}$, thus helping the AAPC$_1$ to move further left on the Nichols plane, as illustrated in Fig. 5.11 for $\omega = 10$. For this purpose, E_{1s} and $|M_1|$ should be kept small, and $|M_2|$ large, i.e., the ratio $|M_2/M_1|$ must be increased. In particular, a linear passive compensator to reduce the typically excessive loop gain in a feedback amplifier in the frequency range $2 < \omega < \omega_b$ must be preferably installed in the link M_1 and not in the link M_2.

5.2.6 Effect of Loop Gain Ignorance

The tolerances of M_1 and M_2 affect the location of the curves in Fig. 5.20. The two most realistic situations are shown in Figs. 5.21 and 5.22.

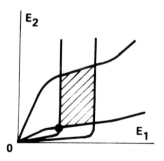

FIG. 5.21

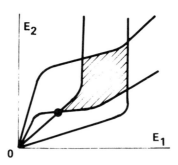

FIG. 5.22

The characteristics of Fig. 5.21 are typical for the band of operation and nearby to it, since at these frequencies the signal in the critical state is limited in the link H_2.

The characteristics shown in Fig. 5.22 are preferred at higher frequencies, outside of the working band, where we may distribute a lesser signal to the input of H_2, and a larger one to the *NE* within the NDC, thus increasing the phase lead produced by the NDC. In this case, the signal that has already been limited in the NDC is too small to saturate the link H_2.

Due to gain ignorance, the intersection of the curves relating to the critical state might fall anywhere within the dashed area. The worst case of the smallest E_{1c} is marked by a bold dot. The worst case corresponds to the smallest phase stability margin, since, as was proved, the dependence arg $H_1(E_1)$ is monotonic. This case corresponds to the smallest values of $|M_1|$ and $|M_2|$, and therefore, $|T_0|$, the latter denoted in the following as $T_{0\,min}$.

Therefore, within the frequency range where $|T_{0\,min}| > x_1$, the phase stability margin need only be checked for the lowest possible gain of the loop links. At higher frequencies, the opposite case of $T_{0\,max}$ should be contemplated.

5.2.7 Sufficient Stability Criterion

Although the number of Bode diagrams that need to be plotted to calculate exactly the frequency response of arg T_c is infinite, we will show in the following that certain practical conditions permit us to estimate the phase stability margin using only a small number of Bode diagrams. This estimation serves as a sufficient stability criterion.

Nonlinear Dynamic Compensator

Generalizing the analysis presented in Sec. 5.2.3, we will consider the feedback system consisting of several nonlinear elements respectively having the DF $w_1, w_2, \ldots, w_j, \ldots$ We presume that across some frequency band the arg T depends monotonically on each of the w_j, and each one of the w_j depends monotonically on the amplitude of the fundamental E_{wj} supplied to this nonlinear element, so that increase in E_{wj} monotonically increases arg T. These two conditions are specifically satisfied if the system is designed according to the recommendations given in Sec. 5.2.4.

Assume that a frequency ω_q within the band at which all E_{wjc} are minimal (subscript c signifies the critical state) is known. For the purposes of analysis, replace the nonlinear elements by their linear equivalents. Denote the return ratio in this linear system as T_{cq}. Then,

$$\arg T_c(j\omega) > \arg T_{cq}(j\omega) \tag{5.22}$$

and arg T_c can be bound from below throughout the considered frequency band by calculation of arg T_{cq} from the Bode diagram for T_{cq}.

The frequency ω_q may be taken as the one at which both the T_0 and the transmission coefficient from the input of the NDC to the NE inside of it accept their smallest values in the considered frequency band. These conditions usually take place at the highest frequency of the band if $|T_0|$ monotonically reduces *versus* ω within this band, i.e., for the majority of applications, throughout the entire frequency range of importance.

Then, a sole Bode diagram with $\omega_q = \{\omega : 20\log|T_0| = x_1\}$ is sufficient to guarantee the system stability in the global sense. The estimated stability margin is smaller than the genuine stability margin, especially in and nearby the band of operation. For better accuracy, the range of importance may be divided into several adjoining subbands, the phase stability margin in each one estimated via the Bode diagram related to the critical state achieved at the highest frequency of the subband.

The typical application is illustrated in Fig. 5.23. We presume that the Bode diagram for $T_0(j\omega)$ lies somewhere within the dashed area, and is parallel to the boundaries denoted as $|T_{0\,min}|$ and $|T_{0\,max}|$, such that the arg T_0 remain unaffected by variations of $|T_0|$.

At frequencies within the interval $[\omega_{q1}; \{\omega : |T_{0\,max}| = x_m\}]$ the phase stability margin exceeds $y\pi$, and no oscillation may occur. Here, x_m serves as the amplitude stability margin for $T_{0\,max}$. It usually comprises about one-third of the amplitude stability margin $x \cong 10\,\text{dB}$ of general use for the average gain system. Then, the stability margin need only be checked out at the frequencies below ω_{q1}.

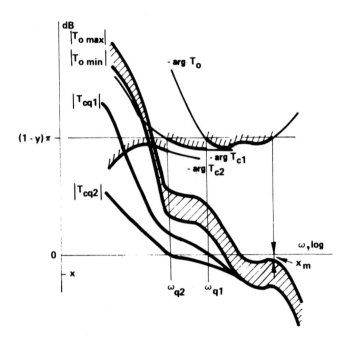

FIG. 5.23

Suppose that a sinusoidal signal with frequency ω_{q1} is applied to the input of the disconnected loop, either to the NDC or to H_2, and the amplitude of the signal is gradually increased until arriving at the condition of critical state. Then, determine the amplitudes of the fundamental applied to each of the *NE*, and the related *NE* DF. Further, replace all the *NE* by equivalent linear elements, yielding a linear circuit having the transfer function $T_{cq1}(j\omega)$. We need only know the modulus of this function to plot the Bode diagram. From this diagram, the arg T_{cq1} can be readily calculated across the frequency range $\omega < \omega_{q1}$.

As demonstrated in Fig. 5.23, $|\arg T_{cq1}| \leq (1-y)\pi$ in the band below ω_{q1} down to a certain frequency ω_{q2}. A periodic self-oscillation within this band cannot exist due to (5.22).

Repeating the procedure with regard to ω_{q2} instead of ω_{q1}, we find that at all frequencies below ω_{q2}:

$$|\arg T_c| < |\arg T_{cq2}| < (1-y)\pi$$

The stability conditions, therefore, are met with the prescribed stability margin at all frequencies, giving us reason to believe that the system is stable in a global sense.

In practice, some of the above steps can be omitted or simplified, especially due to circumstances discussed at the end of Sec. 5.2.3, and the procedure appears to be quite handy.

5.3 NDC IN THE INTERSTAGE CIRCUIT OF A WIDEBAND FEEDBACK AMPLIFIER

The NDC considered in this section serves as the two-pole load for the penultimate stage of a feedback amplifier. Such a NDC must incorporate stray interstage capacitance, and in the linear state of operation the gain of the stage must be maximized. Thus, the maximum of the NDC admittance must be minimized within the frequency band of operation.

As the basis for this design, a variable symmetrical equalizer, described in Sec. 1.4.3, was selected, with the variable resistor replaced by a two-pole NE having a relay-type characteristic $u = u_s$ sign i shown in Fig. 5.24, where u_s stands for the threshold. Such a NE is realizable as a back-parallel connection of two diodes.

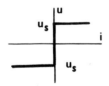

FIG. 5.24

Assume that excitation of the stage is sinusoidal, and the current $i(t)$ passing through the NE is approximately sinusoidal. If so, the DF admittance of the NE is

$$w = \pi I/4 u_s \cong I/1.27 u_s \tag{5.23}$$

When the signal amplitude is small enough, the NE does not conduct; then, $w = \infty$, and from (1.54) and (1.55),

$$H_1 = W(w)/W(\infty) = (wQ + 1)/(w/Q + 1) \tag{5.24}$$

Fairly good results are available with the L-type structure of the auxiliary four-pole. In a NDC such as shown in Fig. 5.25, the amplitude of the current through the NE is

$$I = I_p |Z_1/(Q + 1/w)| \tag{5.25}$$

where I_p stands for the amplitude of sinusoidal current supplied by the amplifier element of the penultimate stage.

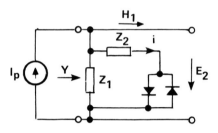

FIG. 5.25

Simultaneously solving (5.23) and (5.25) yields

$$w = \frac{|Z_1 I_p/1.27 u_s|^2 - 1}{\{[|Z_1 I_p/1.27 u_s|^2 - 1]|Q|^2 + \operatorname{Re} Q\}^{1/2} + \operatorname{Re} Q} \tag{5.26}$$

The nonlinearity of the ultimate stage (following the NDC) is assumed to be memoryless saturation. Denoting the threshold of this stage as e_s, the normalized transfer DF of the ultimate stage, in accordance with (3.23), is 1 when $E \leq e_s$ and

$$H_2 = 1.27 e_s/E_2 - 0.27(e_2/E_2)^{-4}: E_2 \geq e_s \tag{5.27}$$

with E_2 the amplitude of the signal at the input of the ultimate stage. The total nonlinearity of the feedback loop described by the DF $H = H_1 H_2$ represents dynamic saturation, so the system jump resonance can be evaluated as described in Sec. 4.3.2.

Next, determine the frequency responses for Z_1 and Z_2. For small amplitude signals, Z_2 is disconnected by the NE, and Z_1 solely plays the role of the load for the penultimate stage. Therefore, $|Z_1|$ must be maximized over the working frequency band under the restriction imposed by the given interstage stray capacitance. The expression for Z_1 and the frequency responses are given by (2.40) and in Fig. 2.48. The responses for the components of Z_2 can be found using (2.40), (1.62), and (1.63). The components of the regulator function $Q = 1/Z_2$ are illustrated in Fig. 5.26, with the boundary frequency $\omega' = 2.13$ determined by the value of regulation at low frequencies $q = 8$ (see Sec. 1.4.3 and Sec. 2.3.3 for definitions).

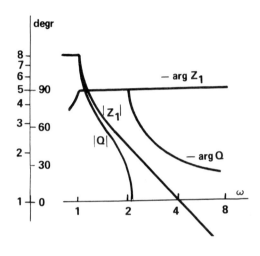

FIG. 5.26

The frequency responses for Z_1 and Q, shown in Fig. 5.26, render a NDC capable of ensuring global stability in the Nyquist-stable system, as we can see from the calculated AAPC in Fig. 5.27. As shown in Fig. 5.28, the small-signal loop gain and phase shift have been constructed as a sum of the Bode optimal cut-off $A_B + jB_B$ and the function $A_1 + jB_1$, where A_1 consists of five constant-slope segments.

The feedback of 51 dB obtained in this Nyquist-stable system is 14 dB greater than that which is available with Bode optimal cut-off (37 dB). To obtain the same (51 dB) feedback in the system with Bode optimal cut-off, the frequency ω_c is required to be 2.6 times higher.

The analysis performed by the author [112, 113] shows that the impact of higher harmonics, neglected in the foregoing consideration, introduces an extra phase lag for the fundamental, reaching 18° at $\omega = 1.4$. This value is considerably smaller than the phase lead of the NDC (of 105° at $\omega = 1.4$) calculated without taking into account the harmonics. As can be seen in Fig. 5.27, the stability margins will remain sufficient with this extra phase lag.

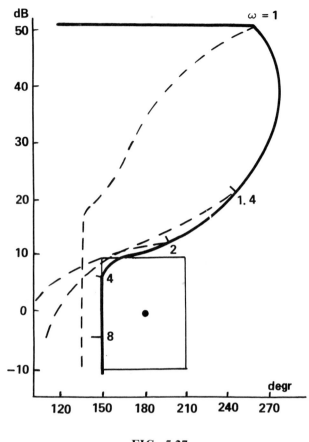

FIG. 5.27

The NDC was designed using the theory of a symmetrical regulator. The symmetry requirement is excessive for the problem at hand. Thus, reconsider the NDC design. Because Z_1 has already been chosen to be optimal for achieving maximum signal gain, it cannot be changed, so it remains for us to find the optimal physically realizable Z_2.

First, prove that with any given w, the phase regulation of the variable equalizer is maximized if

$$\arg Z_2 = \pi/2 \qquad (5.28)$$

$$Z_2 = |Z_1|/2 \qquad (5.29)$$

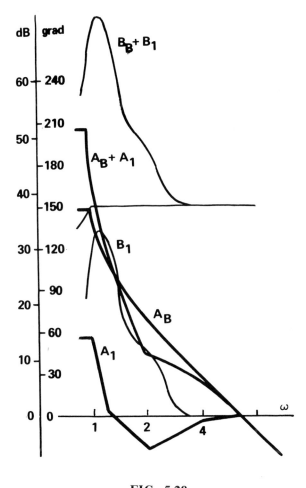

FIG. 5.28

If (5.28) holds, the driving-point admittance for the circuit of Fig. 5.25 is

$$Y = \frac{j}{|Z_1|} + \frac{1}{j|Z_2| + 1/w} = \frac{|Z_1| - |Z_2| + j/w)}{|Z_1| (j|Z_2| + 1/w)} \qquad (5.30)$$

Its phase angle:

$$\arg Y = \arctan[1/w \, (|Z_1| - |Z_2|)] - \arctan(|Z_2|/w) \qquad (5.31)$$

has an extremum, which is, in fact, a minimum, when

$$\frac{\partial \arg Y}{\partial |Z_2|} = \frac{1}{w} \left(\frac{1}{(|Z_1| - |Z_2|)^2 + w^{-2}} - \frac{1}{|Z_2|^2 + w^{-2}} \right) = 0$$

from which (5.29) follows.

On the other hand, with the condition (5.29) fulfilled, from (5.31), the derivative is

$$\frac{\partial \arg Y}{\partial (1/w)} > 0$$

Hence, minimum of arg Y is achieved with the minimum possible resistance in the branch having the impedance $Z_2 + w$, from which the applicability of condition (5.28) follows. Together, therefore, conditions (5.28) and (5.29) minimize arg Y.

However, the conditions (5.28) and (5.29) cannot be simultaneously met throughout the entire desired frequency range because this Z_2 is not physically realizable. The function Z_2, which was used before in the symmetrical regulator, is purely reactive in the most important frequency band $[1, \omega']$, as suggested by (5.28), but its modulus is much smaller than that required by (5.29), as shown by line 1 in Fig. 5.29, and reaches the optimal value of $|Z_1|/2$ only at the frequency ω'. The maximum realizable Z_2 certainly relates to $Z_2 = j\omega/\omega'$, as shown by line 2.

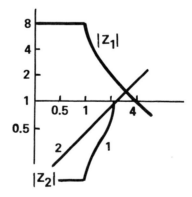

FIG. 5.29

With this Z_2, keeping the notation $Q = Z_1 + Z_2$ (which is, however, no longer the function of regulation), we can find w from (5.26). In the range of importance, where both Z_1 and Z_2 are purely reactive, and for the value of the threshold $u_s = 0.8$, formula (5.26) reduces to

$$w = (I_p|Z_1|^2 - 1)^{1/2}/Q$$

The DF H_1 of the NDC is the ratio of $\{Y: w = 0\}$ to Y,

$$H_1 = (Q + 1/w)/(Z_2 + 1/w)$$

As calculations show [113], such implementation of the NDC increases the minimal phase stability margin (at $\omega = 1.4$) by $18°$.

The two-pole NDC can be employed not only in the interstage network, but also as a parallel local feedback path for the penultimate stage. The element values for this NDC can be calculated using the "Miller effect" relation (according to which a two-pole with impedance Z connecting the input and output of the amplifier, having voltage gain coefficient K, produces the same effect on the transmission function as a two-pole with the impedance $Z/(1 + K)$ shunting its input).

When the NDC is placed in the interstage circuit, the penultimate stage must be capable of supplying the NDC with the much larger signal amplitudes than the threshold of the ultimate stage. Moving the NDC into a local feedback path eliminates this requirement. Since the modulus of the DF of this NDC admittance is similar to dead zone, the DF of the penultimate stage gain represents dynamic saturation. After reaching the threshold of the diodes in the NDC, the output amplitude of the penultimate stage stops growing. It is sufficient, therefore, for its undistorted output signal amplitude to be only 1.2 to 2 times larger than the input threshold of the last stage.

5.4 EXPERIMENTS

Experiment 1. In the Nyquist-stable experimental feedback amplifier diagrammed in Fig. 5.30 [113, 124], the two-pole NDC plays the role of a load for the penultimate stage. Figure 5.31 shows the theoretical (dashed lines) and measured (solid line) frequency responses of the return ratio. Two opposite-biased, back-parallel diodes constitute the *NE*.

Figure 5.32 presents the AAPC for the system without the diodes. Two AAPC, relayed to the frequencies of 65 kHz and 80 kHz, pass through the critical point. Thus, the system is conditionally stable. The oscillation, having the fundamental of 65 kHz, can be incited by the input of a burst of periodic signal, which has its fundamental fairly close to 65 kHz.

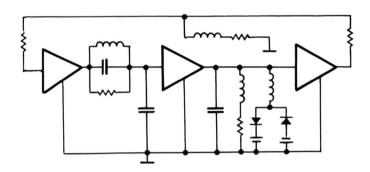

FIG. 5.30

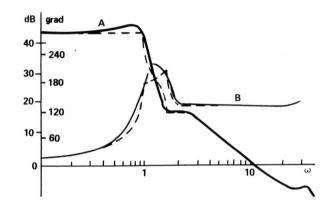

FIG. 5.31

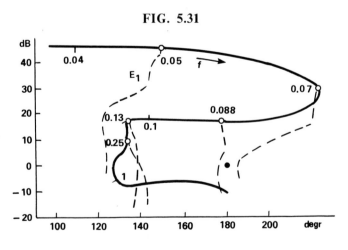

FIG. 5.32

The Nyquist diagram and the AAPC in the presence of the diodes are shown in Fig. 5.33. The two AAPC avoid the critical point. The system remained stable after all attempts to incite self-oscillation by applying signals of various shapes, amplitudes, and frequencies to the system input. When conditions for self-oscillation were created by changing the system configuration, the oscillation vanished as soon as the original system configuration was restored.

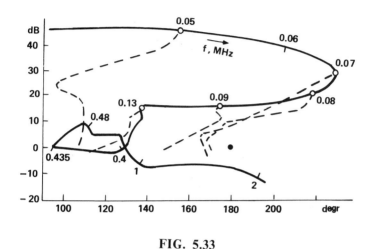

FIG. 5.33

The mouse-shaped loop on the APC, caused by the resonance of the diodes capacitance with the series inductance, was later removed by modifying the NDC as shown in Fig. 5.34(a); the capacitance included in the two-pole realizing the impedance Z_1.

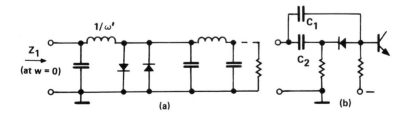

FIG. 5.34

The phase lead produced by the NDC is rather small because the NDC is shunted by the input impedance of the ultimate stage. This shunting can be eliminated using an extra diode, as shown in Fig. 5.34(b), which also serves as an element of the transistor biasing circuit. The blocking capacitor C_2 is large, presenting zero impedance for the ac signal. Combination of small capacitor C_1 with this extra diode forms a second NDC, increasing the total phase lead for the fundamental of large-level signals.

Experiment 2. Linear amplifiers employed in repeaters of a 1300 telephone-channel telecommunication system [113, 126] are Nyquist-stable. Due to larger available feedback, the nonlinear distortions were reduced 10 times as compared with conventional design.

The Bode diagram for the linear amplifiers shown in Fig. 5.35, curve 1, is steep, and $|\arg T|$ exceeds π both at higher and lower frequencies. The NDC was implemented as a parallel local feedback path in the penultimate stage, as shown in Fig. 5.36.

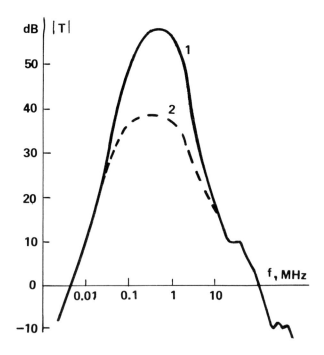

FIG. 5.35

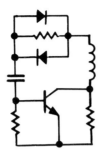

FIG. 5.36

The NDC consists of a pair of back-parallel diodes connected in series with the resonance contour tuned at the center frequency of the working band. As soon as the output signal amplitude of this stage exceeds the threshold of the ultimate stage, the diodes start to open, introducing local negative feedback within and close to the working band. Due to this feedback, the gain around the main loop changes. Curve 2 of Fig. 5.35 shows the main-loop Bode diagram, which, for simulation purposes, was measured with a resistor of 100 Ω in place of the diodes (as if the diodes were open). The system global stability was established using the methods of Sec. 5.2.7.

With the diodes removed, the amplifier is conditionally stable. It is sufficient to touch the interstage circuit elements, or temporarily apply a large signal to the amplifier input, in order to drive the amplifier into self-oscillation. On connecting the diodes, self-oscillation disappears. The amplifiers were manufactured and successfully employed in large quantities.

5.5 SPECIAL APPLICATIONS

1. Even in systems where the available feedback with Bode cut-off is sufficient, using a NDC may be beneficial because it allows larger tolerances for implementation of the frequency response of T. This could reduce the system complexity and the volume of computation necessary in the digital controller.

2. Calculation and computer modeling of complicated feedback systems do not always guarantee stability because of incomplete information regarding the plant parameters. This often necessitates building, testing, and tuning a real-life model. During such tuning, employing a NDC provides AGS. Thus, employment of a NDC may facilitate the design process, even if the NDC will be omitted in the final version of the system.

3. While regulating the system gain by changing the values of $|\beta|$, certain precautions need to be taken in Nyquist-stable systems with NDC. As opposed to the system with optimal Bode cut-off, where reducing $|\beta|$ from the maximum allowable value creates no stability problems, a decrease of $|\beta|$ applied equally at all frequencies reduces the upper amplitude stability margin in a Nyquist-stable system, or may even cause the Nyquist diagram to encompass the critical point. Stability can be retained by increasing the upper amplitude stability margin, but only at the price of reducing the available feedback. When, however, the regulation in β is limited to the working band only by some frequency-selective circuitry, reducing $|\beta|$ makes the Bode diagram less steep, and the stability margins may actually increase.

Chapter 6
Linear Multiloop System

6.1 GENERALITIES

The single-loop feedback system includes either only one amplifier (actuator), or many in a cascade connection so that their transfer coefficients enter the return ratio as multipliers. All of the feedback effects vanish in a single-loop system when any one of the amplifiers is switched off. Therefore, a system with a number of parallel paths through linear circuitry from the output to the input of the string of amplifiers is a single-loop system.

A feedback system that is not single-loop is considered to be multiloop [23]. In fact, a system with several nonlinear links where some are not connected in cascade is considered as multiloop. This definition is unrelated to the number of the system's inputs and outputs, i.e., whether the system is single-input/single-output (SISO) or multi-input/multi-output (MIMO), although MIMO usually is a multiloop system. A SISO system can be either single-loop or multiloop. Although the synthesis theory for the multiloop system is far from complete, many design examples have been demonstrated where multiloop feedback in a SISO system renders better performance than in the single-loop design [11, 16, 18, 23, 52, 58, 59, 65, 89, 101, 162, 176].

Multiloop feedback is widely used in amplification techniques, active circuit synthesis, and automatic control. In wideband feedback amplifiers, the most popular versions of multichannel feedback are emitter feedback (also called cathode feedback [23]) and the combination of the main feedback loop, common for all amplification stages, with the complimentary local loops, each for one or two stages. Various types of multiloop feedback applied in industrial regulators follow diverse properties of the plants under control, which may include cascaded or otherwise connected links with large time delays and large parameter uncertainties.

It is difficult and frequently quite confusing to compare the multiloop feedback design methods described in the literature, especially because of dissimilar sets of initial assumptions, idealizations, and figures of merit. In the following, we will confine ourselves to the study of several multiloop SISO systems whose superiority in available feedback over single-loop systems can be clearly shown.

When the error amplitudes at the inputs to the actuators are normally below the thresholds of saturation, the system may be viewed as linear and, hence, designed by using mathematical methods developed for linear systems; but we must ensure that clipping of the spontaneous large-amplitude input signals in the actuators does not ignite an oscillation or cause large windup. Hence, a viable synthesis theory for "linear" feedback systems should incorporate evaluation of the saturation effects in the actuators. To satisfy this requirement, the design procedures for multiloop systems must be more elaborate than those of single-loop systems described in Chapter 2, where the global stability requirements are formulated rather simply in terms of frequency response of the linear part and, consequently, the system design is reduced to the design of the system's linear part.

In some multiloop systems, the effect of signal clipping in the active links may be adequately approximated by the link gain reduction, similarly to the conventional design of a single-loop system. In this way, real-life nonlinear systems are examined as quasilinear.

6.2 STABILITY CRITERIA

6.2.1 Generalization of the Nyquist Criterion

In application to multiloop feedback systems, the Nyquist stability criterion can be formulated in a variety of ways, depending on the function, or set of functions, which are chosen to be plotted. Among such versions, not many possess the features of simply formulated and well substantiated stability margins; transparency in considering the design trade-offs; and the possibility of at least crudely taking into account the nonlinear effects.

Two different types of Nyquist criteria generalizations are especially important, each one having specific application areas.

The criterion suggested by Bode [23] considers a linear system composed of passive elements and n unidirectional amplifiers. The criterion employs the Nyquist diagrams for the return ratios of the amplifiers T_1, T_2, T_3, ..., T_n plotted with a successively increasing number of amplifiers switched on. The system is stable if and only if the total number of clockwise and counterclockwise encirclements of the critical point -1 are equal for the series of n such diagrams. If the Nyquist diagrams are plotted for return differences $F_i = T_i + 1$, then, certainly, the origin serves as the critical point.

The final result of the stability analysis remains the same with any order of switching on the amplifiers, although the shape of the Nyquist diagrams depends on the order, and proper selection of the order appreciably simplifies the analysis. In particular, it is worth starting with that amplifier which we know does not cause the system to oscillate. This permits us to discard the first Nyquist diagram, since we already know that it does not encompass the critical point.

The Bode-Nyquist multiloop stability criterion comes in handy for application to systems having several local loops and one common loop. Consider, for example, a system with a common feedback path and local feedback around the ultimate stage as shown in Fig. 6.1, which is apparently stable when only the first stage, or only the ultimate stage, is on. This case is quite common due to smaller phase lags in local loops. The stability of the system with both of the amplifiers operating is, therefore, seen from a single Nyquist diagram for either one of the return ratios, for the first or for the ultimate stage, i.e., using either one of the loop cross sections (1) or (2) shown in Fig. 6.1.

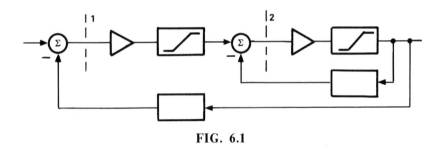

FIG. 6.1

Which diagram to use must be decided in order to simplify the nonlinear analysis. To consider only the effect of the ultimate stage saturation, the cross section (2) is preferred because the DF of the ultimate stage transmission enters this loop return ratio as a multiplier. On the other hand, if the first (penultimate) stage has only a small margin in saturation threshold over the ultimate stage, the maximal output signal from the first stage is too small to change the DF of the ultimate stage noticeably. In this case, the first stage may be considered as the only nonlinear link in the system. Accordingly, choosing the cross section (1) facilitates the analysis.

Using the return ratios for two-poles, the Bode-Nyquist criterion could be applied as well to systems with active or nonlinear two-poles.

Instead of the series of Nyquist diagrams, a single diagram, for the product $F_1 F_2 F_3 \ldots F_n$, is sufficient to establish the fact of system stability. Using a single diagram seems to have an advantage. However, such a diagram does not allow for estimation of the effects of tolerances and nonlinearity of the system links with the convenience of using the series of elementary diagrams.

In the usual MIMO system design, neither of the loops may be considered as the main loop. In this case, generalization of the Nyquist criterion in matrix form [74, 128, 129, 130] conveniently establishes stability margins in terms of allowable tolerances of the system links' transmission functions. Decoupling of elementary loops, i.e., diagonalizing the matrices, is often possible, although requiring extra computations. Without decoupling, the effects of saturation are more difficult to study, and overloading an actuator in one of the loops typically creates significant changes in the return ratios for adjacent loops. This drawback may be reduced or eliminated by installation of NDCs at the inputs to all or some of the actuators.

6.2.2 Examples and Exercises

1. In the feedback system shown in Fig. 6.2, two local positive-feedback loops are added to the common negative feedback. Local feedback increases the gain of the first and third links by 20 dB, thus magnifying the main-loop gain by 40 dB. Hence, the positive local feedback reduces the sensitivity of the system transmission function to variations of μ_1 and μ_3 by 20 dB, and the sensitivity to variations in μ_2 by 40 dB.

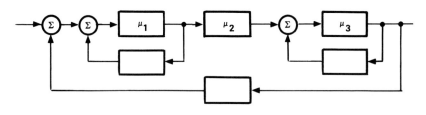

FIG. 6.2

Note that the local feedback may increase the main-loop gain in the working band, but not at the asymptotic frequencies (see further in Sec. 7.1.2). Therefore, application of positive local feedback may benefit the system where the main-loop gain is limited by insufficient gain in the working band, but not by the stability conditions. The drawback of this circuit is that rather small changes in μ_1 and μ_2 would, due to positive feedback, dramatically change the gain of the looped links and the sensitivities.

2. In a three-stage feedback amplifier shown in Fig. 6.3, a subsidiary feedback path around the two first stages introduces positive feedback within the operational frequency band [18], thus increasing the gain over the common loop for better damping of the nonlinear distortions originated in the ultimate stage.

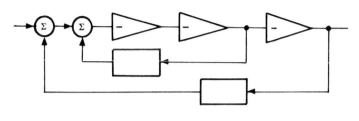

FIG. 6.3

Using the multiloop Bode-Nyquist criterion, determine how the system stability can be assured in the case when the first two stages are unstable with their feedback path. What happens if (a) the ultimate stage is overloaded, or (b) when the clipping in the second stage appears simultaneously with that in the ultimate stage (due to appropriate threshold adjustment)?

6.3 FEED-FORWARD

As a result of (1.47) and (1.52), the sensitivity of W to variations of a parameter w vanishes if and only if $W(w) = W(0)$. Suppose W is understood to be the system transmission function, and w is the amplification coefficient of an amplifier. Then, the sensitivity reduces to zero if and only if switching off the amplifier does not change the system transmission function. This explicitly shows the interreaction between the two issues of sensitivity reduction and of creating the system redundancy by organizing at least two transmission channels, which forward the signal from the input to the output.

H.S. Black's idea of feed-forward compensation [24] is illustrated in Fig. 6.4. The system transmission coefficient is $\mu_1 + \mu_2 - \mu_1\mu_2\beta$. It reduces to $1/\beta$ if either μ_1 or μ_2 equals $1/\beta$, regardless of the transmission coefficient of the remaining amplifier. Thus, if transmission coefficients of the amplifiers are

$$\mu_1 = \mu_2 = 1/\beta \tag{6.1}$$

the sensitivity of W to either μ_1 or μ_2 drops to 0.

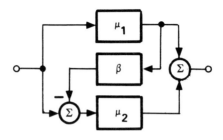

FIG. 6.4

The delightful result of zero sensitivity (and zero nonlinear distortion) relates, however, only to the case described by (6.1). When μ_1 and μ_2 deflect from the value $1/\beta$, the sensitivities drastically rise. To keep the tolerances of μ_1 and μ_2 small, it is, therefore, advantageous to implement large internal negative feedback in each of the amplifiers [135, 176] as shown in Fig. 6.5.

From the diagram of Fig. 6.5(a), it follows immediately that

$$W = \frac{U_2}{U_1} = \frac{\mu_1}{1 + \mu_1 \beta_1} + \frac{\mu_2}{1 + \mu_2 \beta_2} - \frac{\mu_1 \mu_2 \beta}{(1 + \mu_1 \beta_1)(1 + \mu_2 \beta_2)}$$

or

$$W = \frac{\mu_1 + \mu_2 + \mu_1 \mu_2 (\beta_1 + \beta_2 - \beta)}{(1 + \mu_1 \beta_1)(1 + \mu_2 \beta_2)} \tag{6.2}$$

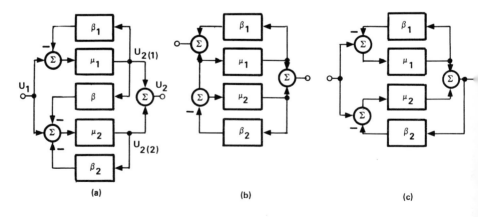

(a)   (b)   (c)

FIG. 6.5

Linear Multiloop System

The sensitivities of W to μ_1 and μ_2, respectively, are given by

$$S_1 = \frac{d \ln W}{d \ln \mu_1} = \frac{1 + \mu_2(\beta_2 - \beta)}{(1 + \mu_1 \beta_1)[1 + \mu_2/\mu_1 + \mu_2(\beta_1 + \beta_2 - \beta)]} \quad (6.3)$$

$$S_2 = \frac{d \ln W}{d \ln \mu_2} = \frac{1 + \mu_1(\beta_1 - \beta)}{(1 + \mu_2 \beta_2)[1 + \mu_1/\mu_2 + \mu_1(\beta_1 + \beta_2 - \beta)]} \quad (6.4)$$

Since the feedback paths could be built of precision passive elements, the equality:

$$\beta_1 = \beta_2 = \beta$$

could be realized, leading to

$$W = \frac{\mu_1 + \mu_2 + \mu_1 \mu_2 \beta}{(1 + \mu_1 \beta)(1 + \mu_2 \beta)} \quad (6.5)$$

and

$$\lim_{\mu_1 \to \infty} W = \lim_{\mu_2 \to \infty} W = 1/\beta$$

The sensitivities are

$$S_1 = \frac{1}{(1 + \mu_1 \beta)(\mu_2/\mu_1 + 1 + \mu_2 \beta)} \quad (6.6)$$

$$S_2 = \frac{1}{(1 + \mu_2 \beta)(\mu_1/\mu_2 + 1 + \mu_1 \beta)} \quad (6.7)$$

When the local feedback is large, i.e., $|\mu_1 \beta| \gg 1, |\mu_2 \beta| \gg 1$, and the values of $|\mu_1|$ and $|\mu_2|$ are of the same order, then, as can be seen from (6.6) and (6.7), the sensitivities are roughly inversely proportional to the product of the return differences for the local loops. Hence, the nonlinear distortions and the gain tolerances are effectively suppressed.

When $\beta_1 = \beta_2 = \beta$, we can easily assure ourselves of the equivalency of the transform of the diagram shown in Fig. 6.5(a) to those shown in Fig. 6.5(b,c) by observing the paths of transmission of the signals to the input of the amplifier μ_2.

The signals at the outputs of the amplifiers μ_1 and μ_2 are given by

$$U_{2(1)} = U_1 \frac{\mu_1}{1 + \mu_1 \beta}$$

and

$$U_{2(2)} = U_2 \frac{\mu_2}{(1 + \mu_1 \beta)(1 + \mu_2 \beta)}$$

When the local feedback is large, $|U_{2(1)}| \gg |U_{2(2)}|$. Therefore, the first amplifier is the main one, determining the system output power capability.

The effect of input noise sources N_1 and N_2 of these amplifiers at the system's output is given by

$$N_1 \frac{\mu_1}{(1 + \mu_1 \beta_1)(1 + \mu_2 \beta_2)} + N_2 \frac{\mu_2}{(1 + \mu_2 \beta_2)} \tag{6.8}$$

When the local feedback is large, the second term in this expression dominates.

Therefore, in order to achieve large dynamic range of the input signals, the available output power of the first amplifier has to be large, and the noise figure of the second amplifier has to be small.

The feed-forward circuit can be implemented in a diversity of ways, for example, as in the amplifiers in Figs. 6.6 and 6.7. Generally, it is not required for the summers to be ideal, and the feed-forward effect can be achieved in the circuit of Fig. 6.8 with passive multipoles at the input and the output, which have zero transmission in the directions shown by crossed dashed lines, so that the signal from the output of the amplifier μ_2 does not arrive at the input of μ_1.

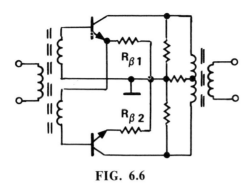

FIG. 6.6

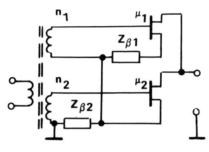

FIG. 6.7

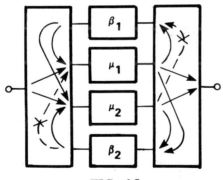

FIG. 6.8

Implementation of feed-forward systems requires precise adjustment of the channel transmission coefficients, which is not convenient in devices intended for mass production. The feed-forward principle is only worth applying when the required small values of the sensitivity cannot be realized with the feedback.

Unlike the feedback system, large NPS in the channels does not preclude implementation of the feed-forward system, only requiring that these phase shifts be made equal. The feed-forward principle was applied, for example, for reduction of nonlinear distortion in traveling wave tube (TWT) amplifiers, where the large NPS makes impossible using the feedback.

An example of a feed-forward broadband amplifier is shown in Fig. 6.9. The system employs two amplifiers, four directional couplers with low attenuation for the forward signal propagation (typically 1 dB) and rather large attenuation in other directions (typically 10 dB), an attenuator, and two delay lines (DL).

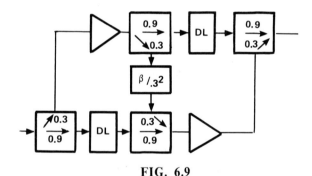

FIG. 6.9

6.4 SYSTEM WITH PARALLEL AMPLIFICATION CHANNELS

6.4.1 Sensitivity

In a single-loop feedback system, the sensitivity is inversely proportional to the feedback. The same conclusion holds for the system of Fig. 6.10 with several parallel amplification channels, provided that these are replaced by an equivalent amplifier with transmission function $\mu = \Sigma_i \mu_i$.

Assume the amplification of the first (main) amplifier $|\mu_1| \gg |\mu_i|$, for all i, in the working band. In this case, evidently, $\mu \cong \mu_1$ and the sensitivity S_1 to μ_1 approaches the sensitivity S to μ, i.e.,

$$S_1 \cong S = 1/(\beta \sum_i \mu_i + 1) = 1/T_{01}$$

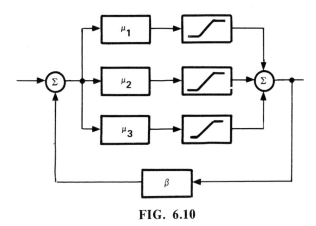

FIG. 6.10

where $T_{01} = -\beta\mu_1$ stands for the return ratio for the first channel measured under the condition that all other channels are cut off.

The sensitivity of the first channel measured while other channels are operating differs from the above value. It may be calculated using (1.10) and (1.33) without neglecting k_d.

6.4.2 Stability

For the system of Fig. 6.10, the fact of stability can be established by observing the shape of the Nyquist diagram for the return ratio of the equivalent amplifier:

$$\mu\beta = \sum_{i=1}^{m} T_{0i}$$

where $T_{0i} = \mu_i\beta$.

In a real system, the ultimate stage of each channel includes a saturation element that reduces its DF for large signals. We assume here that the filtering properties of the channels permit using these DF values for the stability analysis.

In the following, we limit our consideration to the system satisfying these conditions:

1. $|\mu_{i-1}| \gg |\mu_i|$, over the operating band, for all i.

2. With gradually increasing the amplitude of the sinusoidal signal at the common input to the channels, limiting starts first in the ith channel and then in the $(i+1)$th channel for all i. Therefore, the maximum output power of the channels can be reduced with i. We will consider further the system designed in exactly this way. The first channel is then the main one, having the largest gain and output power capability.
3. With the signal level reaching the threshold of saturation in the ith channel, all the channels with numbers less than i are considered to be not operating, either because of saturation, or, if this is not enough, as being switched off by special logic circuitry.

Under these circumstances, for any given amplitude of the sinusoidal input signal to the channels, only the channels with numbers $i, i+1, \ldots, m$ are operative, and only the ith may have the gain reduced as a result of limiting. Such a system can be easily realized because the threshold of the controller that should govern switching the channels off can be chosen much larger than the threshold of saturation in the main channel.

For this system to be stable, it is necessary and sufficient that for any i and any $H_i \in [0,1]$, the Nyquist diagram for $H_i T_i + T_{i-1} + \ldots + T_1 + 1$ does not encircle the critical point $(0,0)$, or, which amounts to the same thing, that the Nyquist diagrams for

$$\frac{H_i T_{0i}}{T_{0\ i-1} + T_{0\ i-2} + \ldots + 1}$$

does not encircle the point $(-1,0)$. Thus, in the latter case, for all i, the plots for

$$\frac{T_{0i}}{T_{0\ i-1} + T_{0\ i-2} + \ldots + 1} \tag{6.9}$$

must not cross the real negative semiaxis to the left of the point $(-1,0)$ on the Nyquist plane, i.e., that they have the shape of a Nyquist diagram in an absolutely stable single-loop system.

Next, we may plot the following sequence of Bode diagrams:

1. The diagram for the channel with the smallest gain T_{01} must guarantee this channel global stability while the others are all switched off. Hence, it has to be a Bode optimal cut-off determined by the given high-frequency asymptote. This diagram is shown by the curves 1 in Fig. 6.11(a) with $x = 8.7\,\text{dB}$, $y = 1/6$, $\omega_a = 5$, $n = 2$. The dashed line represents the phase shift. Curves 2 relay to $T_{01} + 1$.

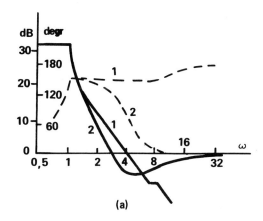

(a)

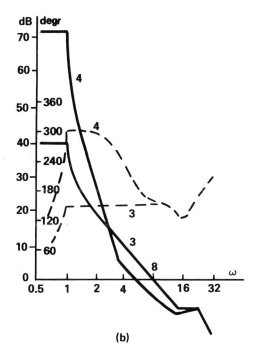

(b)

FIG. 6.11

2. With the mth and $(m-1)$th channels on, the return ratio for the next $(m+1)$th channel is

$$\frac{T_{0\,m+1}}{T_{0m}+1} \qquad (6.10)$$

At high frequencies this function approaches T_{0m-1}. Therefore, the Bode diagram for (6.10) may be implemented as the Bode optimal cut-off with the same position of the asymptote as for T_{0m-1}, as shown by the curve 3 in Fig. 6.11(b). Here, the stability margins are $x = 8.7\,\text{dB}$, $y = 1/6$, and the parameters of the high frequency asymptote $n_{m-1} = 3$ and $\omega_{a\,m-1} = 16$.

The Bode plot for $T_{0\,m-1}$ shown in Fig. 6.11(b), the curves 4, is found by summing the responses 3 and 2.

The feedback in the two-channel system is great, which readily suffices for practical needs, so the number of channels need not be increased. Theoretically, however, we may proceed further.

3. If the system contains a third channel (not shown in Fig. 6.11), its return ratio, which is found as

$$\frac{T_{0\,m-3}}{T_{0\,m-1}+(T_{0m}+1)} \qquad (6.11)$$

approaches $T_{0\,m-3}$ at higher frequencies. Then, the function (6.11) can be realized as the Bode optimal cut-off, and the related frequency response for $T_{0\,m-3}$ can be found.

The frequency characteristics for the fourth channel, *et cetera*, may be calculated similarly. The overall attainable feedback in the operational band equals in logarithmic units the sum of the amounts attainable in each channel taken separately.

In an example with two amplification channels having the responses given in Fig. 6.11, the total feedback is 72 dB as opposed to 40 dB in the single-channel system.

The calculated Nyquist diagrams for $T_{02} + H_1 T_{01}$ are displayed on the Nichols plane in Fig. 6.12. It can be seen that although decreasing H_1 changes the shape of the diagrams, all the diagrams avoid the critical point. The shape of the diagram for $H_1 = 0$ guarantees absolute stability with respect to H_2. Hence, the system also remains stable under any degree of overloading the link H_2.

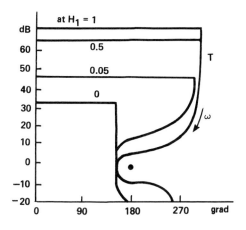

FIG. 6.12

The described method of calculating the responses for the sum $T_1 + T_{02}$ *et cetera* may be employed to generate frequency responses for Nyquist-stable systems.

Chapter 7
Nonlinear Multiloop System: Describing Function Approach

7.1 LOCAL FEEDBACK

7.1.1 Cascaded Links

Local feedback is suggested to supplement common feedback when the cascaded links constituting the plant widely differ in their frequency characteristics and in the required feedback. Suppose, for example, that an operational amplifier with wider gain bandwidth, but larger gain tolerances, is placed in the common loop ahead of relatively slow and accurate electromechanical links. By adding a local loop for the amplifier, the total output error appreciably decreases.

Finding the best possible trade-off between the values of common and local feedback might be complicated, encumbered by the fact that introduction of a local linear feedback path converts a nondynamic nonlinear link into a dynamic nonlinear link, so that overloading in this link changes the common-loop phase lag and the maximum available common feedback.

When the phase stability margin in a local loop is small, jump-resonance makes the link's DF multivalued. The stability analysis of a system including such a link must examine all the choices for partial solutions.

Many other feedback configurations, such as crossed feedback loops, can be analyzed through equivalent transformation into systems employing local and common feedback loops.

7.1.2 The Feedback Around the Ultimate Stage

Consider the feedback applied to reduce nonlinear distortions in an amplifier whose ultimate stage contributes the dominant part of these distortions. Assume that the available common feedback is insufficient, limited by the loop-gain degradation at higher frequencies.

A local loop may be added to the common loop either around the stages of preliminary amplification, as shown in Fig. 7.1, or around the ultimate stage (plant), as shown in Fig. 7.2.

If the output power capability of the penultimate stage is made large in order to reduce its nonlinear distortions, the stage can be considered as linear. Then, the frequency response of the plant return ratio must be implemented as the Bode optimal cut-off. The local feedback path may increase the available feedback if it improves the high-frequency gain asymptote around the plant.

This improvement can be attained in the system of Fig. 7.1 in two different manners. First, by making the local feedback positive in the operational band to increase the gain $|\mu_1/(1-\mu_1\beta)|$. Such a practice benefits, for example, the design of a multistage feedback amplifier with large external gain, where the local positive feedback allows for fewer stages, thus making the asymptotic losses smaller.

Secondly, the asymptotic losses may be reduced by the local feedback if the feedback is positive in the frequency region of Bode step on the Bode diagram for the common loop. Simultaneously, negative feedback appears at lower or higher frequencies because the feedback integral (2.11) equals zero. Detailed analysis shows that the increment in the plant feedback achieved this way is rather small: less than 3 dB [108, 113].

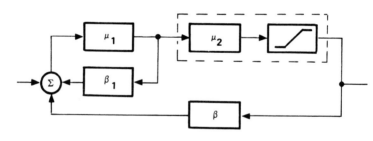

FIG. 7.1

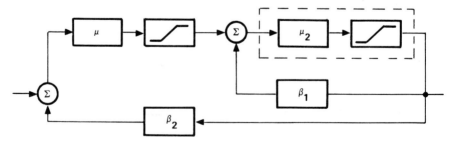

FIG. 7.2

Next, consider the block diagram of Fig. 7.2, which is equivalent in the linear state of operation to the single-loop system of Fig. 7.3. In the single-loop system, saturation of the penultimate stage does not affect the stability condition and the available common feedback. In the system of Fig. 7.2, this saturation link benefits the system stability, as can be seen from the following.

In the band of operation and within 1 or 2 octaves beyond, the phase lag around the local loop is smaller than that around the common loop. Therefore, the signal with an amplitude large enough to be limited in the penultimate stage changes the ratio in amplitudes of the fundamentals of the two signals attending the combiner. The phase lag of the total signal becomes lower, and the AAPC T avoids the critical point by a greater distance. This is the reason that the block diagram of Fig. 7.2 is widely used in automatic control and feedback amplifiers design.* For example, it is used in aircraft control [18] with the plant representing the dynamics of the aircraft and the servo, $\mu_2 = 1$, and μ_1, which is a differentiator, a sensor of the angle velocity.

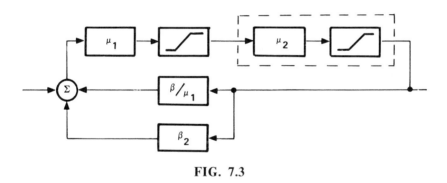

FIG. 7.3

It is worth making the threshold of saturation in the penultimate stage frequency-dependent by introducing a linear compensator with the attenuation increasing at the frequencies just beyond the working band (recall Sec. 2.43). This allows for larger phase lag and increased slope of the Bode diagram in this frequency region.

As an example of more complex interaction between the loops, consider emitter feedback in a feedback amplifier, where the collector current of the transistor represents the system output, and the return signal is proportional

---

*The explanation in the literature of why such a system is introduced is often based on only analysis of the linear (small-signal) state of operation, and is not quite correct.

to the emitter current $I_e = I_c + I_b$ as shown in Fig. 7.4(a). The phase lag of I_b is smaller than that of I_c. The equivalent block diagram of Fig. 7.4(b) comprises symmetrical limiting for I_c and the one-sided limiting for I_b. When the voltage at the stage input grows larger, the AAPC T traverses to the left, away from the critical point. This and the improved high-frequency asymptote of the return ratio are the advantages of the emitter feedback. The improved high-frequency asymptote is due to lesser phase lag in the transistor transmission function in the direction basis-emitter than in the direction basis-collector, and in smaller stray capacitance shunting the feedback path, as illustrated in Fig. 7.5.

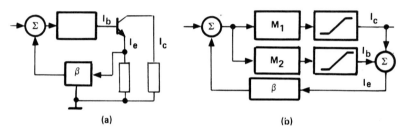

FIG. 7.4

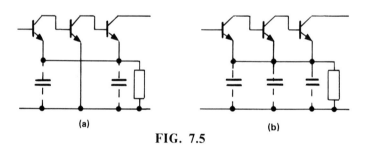

FIG. 7.5

The disadvantage of emitter feedback is that the current I_e and not the output signal $I_c = I_e - I_b$ is stabilized and linearized. The current I_b contributes substantial error in precision amplifiers, especially at higher frequencies, where I_b is relatively large and contains nonlinear products [103].

Using frequency-selective circuits, as in the case of Fig. 7.6 with voltage divider Z_1, Z_2, the $|Z_1|$ being small in the working band and $|Z_2|$ at higher frequencies, we can combine the advantages of the emitter and single-loop feedback [101, 103].

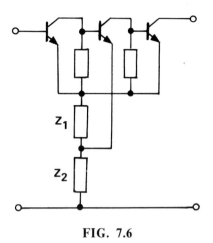

FIG. 7.6

7.2 TWO PARALLEL CHANNELS WITH SATURATION

7.2.1 Frequency Responses of the Linear Part

Presume that for the system of Fig. 7.7 $|T_{01}| = |\beta M_1| \gg |\beta M_2| = |T_{02}|$ over the band of operation, i.e., the main channel is the first one. The following design procedure is aimed at maximization of the feedback for the total equivalent amplifier $|T_{01} + T_{02} + 1| \cong |T_{01}|$ over the operational band while ensuring the system global stability.

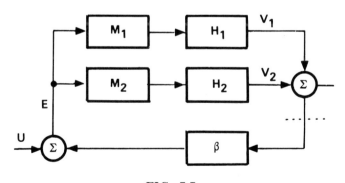

FIG. 7.7

We will use the values of the signal normalized with respect to the threshold of saturation of the main channel. Denote the amplitude of the system's input fundamental as U, the fundamental's amplitude at the input of the channels as E, at their outputs, respectively, as V_1 and V_2; the maximal amplitude of the fundamental at the channel outputs as V_{S1} and V_{S2}; the maximum instant amplitudes at the channel outputs as v_{S1} and v_{S2}; the maximal amplitude E at which saturation started in the second (auxiliary) channel, as e_{S2}. Then,

$$\frac{v_{S2}}{v_{S1}} = \frac{V_{S2}}{V_{S1}} = e_{S2} \frac{|T_{02}|}{|T_{01}|}$$

Practical considerations restrict the available power from the second channel output, i.e., v_{S2}. If, however, v_{S2} is excessively small, the auxiliary channel does not improve the system stability.

To ensure global stability with minimum possible v_{S2}, the locus $T_{01} + T_{02}$ is required to pass through the vertices k, l, and m of the sector of stability margins, as shown in Fig. 7.8, at the frequencies respectively denoted as ω_k, ω_l, ω_m, and the AAPC $T_1 + T_2$ should pass close to the boundary of this sector.

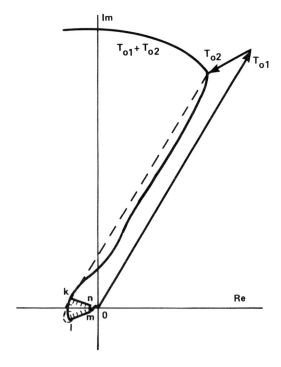

FIG. 7.8

We will analyze the AAPC in the frequency region $\omega \in (1, \omega_l)$, where the stability conditions are critical; if they are satisfied, those AAPC at other frequencies avoid the stability margin automatically, without special care being taken.

Let us find the frequency responses for T_{01} and T_{02}. At $\omega > \omega_l$, i.e., for $|T_{01} + T_{02}| < 10^{x/20}$, the frequency responses for T_{01} and T_{02} must be implemented as Bode cut-offs, so that $\arg T_{01} \cong \arg(T_{01} + T_{02}) \cong (1-y)\pi$. We choose the amplitude stability margin in each channel to be $x = 6\,\text{dB}$, as shown in Fig. 7.9, in order for $20\log|T_{01} + T_{02}| < x$ in the region of the Bode step.

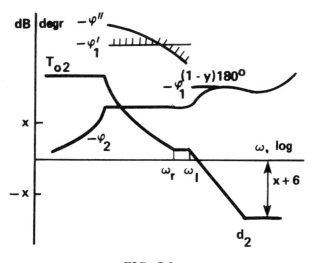

FIG. 7.9

The asymptote of each channel transmission being given, the frequencies ω_{d1} and ω_{d2} can be determined as in Sec. 2.4.1. Further, since at $\omega = \omega_l$,

$$\left. \begin{array}{l} |T_0(j\omega_l)| = |T_{01}(j\omega_l)| + |T_{02}(j\omega_l)| = 10^{x/20} \\ |T_{01}(j\omega_l)| / |T_{02}(j\omega_l)| = (\omega_{d1}/\omega_{d2})^{2/(1-y)} \end{array} \right\} \quad (7.1)$$

we can find $|T_{01}(j\omega_l) + T_{02}(j\omega_l)|$. The frequency:

$$\omega_l = \omega_{d1} \left(|T_{01}| 10^{(x+6)/20} \right)^{(1-y)/2} \quad (7.2)$$

can be calculated or determined graphically from Fig. 7.9.

It will be seen in the following that at $\omega < \omega_l$, the phase stability margin in the auxiliary channel y_2 must be larger than y. For this reason, the Bode plot for T_{02} should have an average slope of $(1 - y_2)$ dB/octave and must possess a step over the interval $[\omega_r, \omega_l]$ with the relative bandwidth

$$\frac{\omega_l}{\omega_r} = \frac{1 - y_2}{1 - y} \tag{7.3}$$

Now, the frequency response for T_{02} is completely defined, as well as for T_{01} at the frequencies $\omega \geq \omega_l$. Let us determine T_{01} at lower frequencies.

At $\omega = \omega_l$, T_{01} must be such that the three following conditions are satisfied. First, as follows from the plot of Fig. 7.8,

$$|T_{01} + T_{02}| \geq 10^{x/20} : \omega < \omega_l \tag{7.4}$$

Second, arg T_{01} is bounded by these inequalities:

$$-\arg T_{01} \leq -\phi_{sens} \tag{7.5}$$

$$-\arg T_{01} \leq -\phi_{(kl)} \tag{7.6}$$

$$-\arg T_{01} \leq -\phi_{(lm)} \tag{7.7}$$

where

$$\phi_{sens} = -\arg T_{02} = (1 - y_2)180° + (1 - y)180° \tag{7.8}$$

in order to delimit the sensitivity of $T_1 + T_2$ to the system's parameter variations; and $\phi_{(kl)}$ and $\phi_{(lm)}$ are determined by the conditions of the AAPC touching the boundary of stability margin as shown in Fig. 7.10 by the broken line, correspondingly, on the arc kl and on the segment lm. These conditions will be studied in the next section.

7.2.2 Piecewise Analysis of the AAPC

We will examine the AAPC piece by piece for

$$T = T_1 + T_2 = T_{01} H_1 + T_{02} H_2 \tag{7.9}$$

shown in Figs. 7.8 and 7.10.

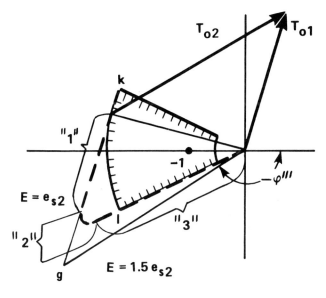

FIG. 7.10

On the first section, for $E \in [1, e_{S2}]$, the DF $H_2 = 1$, thus,

$$T = T_{01} H + T_{02}$$

Increasing E causes the DF H_1 to increase. Then, at this section, the AAPC progresses along a straight line in parallel with the phasor T_{01}.

At frequencies $\omega \in [1, \omega_k]$, for the inequality (7.6) to be satisfied at its limit, the AAPC must pass through the vertex k, as illustrated in Fig. 7.8.

At frequencies $\omega \in [\omega_k, \omega_l]$, the AAPC must be tangent to the arc kl, as shown in Fig. 7.10, and the phase shift $-\phi_{(kl)}$ decreases monotonically with the tangency point tending toward l, i.e., with the frequency increasing.

It follows from the analysis of the triangles of Figs. 7.8 and 7.10 that

$$-\phi_{(kl)} = \begin{cases} (1+y)180° + \arctan \dfrac{\sin[(1+y)180° - \phi_2]}{10^{x/20}/|T_{02}| - \cos[(1+y)180° - \phi_2]} \\ \quad : -\arg(T_{01} + T_{02}) > -(1.5+y)180°, \text{ i.e., } \omega \in [1, \omega_k] \\ (1-y)180° + 90° : -\arg(T_{01} + T_{02}) = -(1+y)180°, \text{ i.e., } \omega = \omega_k \\ (1-y)180° + 90° + \arccos(10^{x/20}/|T_{02}|) : -\arg(T_{01} + T_{02}) \in \\ \quad [(1-y)180°, (1+y)180°], \text{ i.e., } \omega \in (\omega_k, \omega_l) \end{cases}$$

(7.10)

The detailed analysis of these formulas shows that $|T_{02}|$ is worth maximizing (as already assumed), since increasing T_{02} at lower frequencies permits $-\phi_{(kl)}$ to attain its maximum at $\omega = 1$.

The second piece of the AAPC, for $E \epsilon [e_{S2}, 1.5 e_{S2}]$, is curvilinear and relatively short. At this section, using (3.25), the return ratio is

$$T = 1.27 T_{01}/E + T_{02} \tag{7.11}$$

The numerical analysis shows that this part of the AAPC always avoids the stability margin if the AAPC does behave in this way at the first and third pieces.

On the third piece of AAPC, $E = 1.5 e_{S2}$; both channels are saturated, so that $V_{S2}/V_{S1} = v_{S2}/v_{S1}$ and the return ratio is

$$T \cong 1.27 T_{01}/E + 1.27 e_{S2} T_{02}/E = 1.27(T_{01} + e_{S2} T_{02})/E \tag{7.12}$$

The increase of E reduces $|T|$ without affecting $\arg T$, and the AAPC represents a segment of a straight line directed toward the origin.

Let us measure T in the cross section shown by the dotted line in Fig. 7.8. The output of the loop is then $\mathbf{V}_{S1} + \mathbf{V}_{S2}$. The vector diagram shown in Fig. 7.11 is similar to that for T (assuming the voltage given to the input of the cross-sectional loop to be negative). Therefore, for the stability margins to be appropriate, the following inequality must hold:

$$-\arg(\mathbf{V}_{S1} + \mathbf{V}_{S2}) \leq (1 - y)\pi \tag{7.13}$$

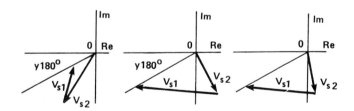

FIG. 7.11

The limiting case of this inequality is illustrated in Fig. 7.1(a). Applying the sine theorem for this triangle (that the ratio of the side of a triangle to the sine of the opposite angle is equal for all the sides), the ratio:

$$\min \frac{V_{S2}}{V_{S1}} = \frac{\sin|\phi_1''' - (1-y)180°|}{\sin|\phi_2 - (1-y)180°|} = \frac{\sin|\phi_1''' - (1-y)180°|}{\sin|y_2 - y_1|180°} \tag{7.14}$$

This dependence is charted in Fig. 7.12 for $y = 1/6$, for the interval of $-\phi_{(lm)}$ from 150° (with lesser $-\phi_1$, the system is AGS without the auxiliary channel) up to the $-\phi_{sens}$, the latter depending on y_2, as seen in (7.8).

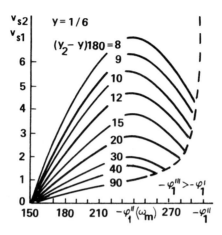

FIG. 7.12

The required v_{s2}/v_{s1} is the largest if $\phi_{(lm)} = (1.5 - y)180°$, i.e., for $\phi_{(lm)} = \phi_{(kl)}(\omega_k)$. Therefore, if the maximal $-\phi_1$ is to be achieved within the limitations imposed by inequalities (7.5) and (7.6) to guarantee

$$-\phi_{(lm)} = -\phi_{(kl)} \tag{7.15}$$

at all frequencies, the strongest requirement for v_{s2}/v_{s1} is imposed by the condition related to the frequency ω_k.

If the problem is stated another way, such that v_{s2}/v_{s1} is prescribed, we have to choose $y_2(\omega_k)$. The smaller ratio v_{s2}/v_{s1} leads to greater y_2, which reduces $-\phi_{sens}$ and, due to smaller related T_{02}, also reduces $-\phi_{(kl)}$.

If the ratio $v_{s2}/v_{s1} < 1$, the inequality (7.15) can no longer be achieved, and just the $\phi_{(kl)}$ delimits the maximum available $-\arg T_{01}$. As is found from (7.14) and illustrated in Fig. 7.11(b), $-\phi_{(kl)}$ is maximal if $y_2 = 0.5 + y$, i.e., $-\phi_2 = (1.5 - y)180°$. However, in this case $|\mathbf{V}_{s2} + \mathbf{V}_{s1}| < V_{s1}$, which reduces $\min(E > 1) M$ and, consequently, due to (4.15), reduces E'_a. The better choice of ϕ_2 leads to the condition $|\mathbf{V}_{s1} + \mathbf{V}_{s2}| = V_{s1}$ illustrated in Fig. 7.11(c) with

$$-\phi_2 = (0.5 - y)180° + \arcsin \frac{V_{s2}}{V_{s1}} \tag{7.16}$$

and with

$$-\phi_1 = (1-y)180° + 2 \arcsin \frac{V_{s2}}{V_{s1}} \quad (7.17)$$

Up to this point, we have assumed the frequency response for T_{02} to be selected as the optimal Bode cut-off. Now, consider the possibilities for the system improvement by variation of this response.

First, $|T_{02}|$ no longer need be kept constant in the band of operation. It may be reduced at lower frequencies, where $\arg|T_{01}|$ is small and the auxiliary channel is not needed, and increased in the neighborhood of $\omega = 1$, in turn increasing $-\phi_{(kl)}(1)$. The numerical analysis shows, however, that the requirements to the values of jumps (of jump-resonance) in the band of operation limit this benefit.

Second, y_2 need not be kept constant within the band $[1, \omega_l]$, but might be increased nearby $\omega = \omega_k$, for example, by making a step on the Bode plot of T_{02} in this region. Due to this fact, the ratio v_{s2}/v_{s1} can be reduced.

7.2.3 Jump-Resonance

Consider two domains of the amplitude E.

For $E < e_{s2}$, the auxiliary channel transmission is linear. Hence, the transmission around the saturation link of the main channel is $T_{01}/(T_{02}+1)$. Using this return ratio, the conditions for the jumps are easy to find from the diagrams of Figs. 4.20 to 4.22.

For $E > 1.5 e_{s2}$, both channels are saturated and the system reduces to an equivalent single-loop system with a sole saturation link looped by the linear link $T_{01} + e_{s2} T_{02} = T_{01}(1 + V_{s2}/V_{s1})$. The diagrams of Fig. 4.22 give the conditions for the jumps.

Within each of these domains on the E-axis (we omit the analysis for the relatively small region $[e_{s2}, 1.5 e_{s2}]$), only one minimum of the input signal amplitude U could take place, as illustrated in Fig. 7.13, E'_a relating to the smallest one.

When $v_{s2}/v_{s1} < 1$ with the conditions (7.16) and (7.17) holding, the modulus of the DF for the composed channel equals H_1. Therefore, E'_a can be found by using the methods developed for the system with dynamic saturation.

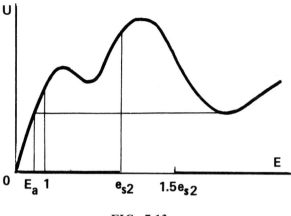

FIG. 7.13

7.2.4 Frequency Response of the Main Channel Transmission Function

The Bode diagram for a physically realizable MP function $T_{01}(j\omega)$ that satisfies the foregoing limitation is conveniently approximated by the sum of three components: the optimal Bode cut-off $A_{(1)}$ found by continuation of the already found (for $\omega = \omega_l$) part of $T_{01}(j\omega)$; the characteristic $A_{(2)}$ of the segment with the slope $12(1-y)\,\text{dB/octave}$; and the real part $A_{(3)}$ of a MP function described by (2.36), in which the imaginary part $B_{(3)}$ remains constant, denoted as B_m, over the interval $[1, \omega']$ as shown in Fig. 7.14..

The condition $B_{(3)} = \phi_1''' - B_{(1)} - B_{(2)}$ applying to the $\omega = 1$ yields B_m. From this condition applied to $\omega \geq \omega_l$, we choose one of the responses $B_{(3)}$ with appropriate ω' among those charted in Fig. 2.43.

The obtained response for T_{01} is fairly close to optimal. It cannot be improved because further increase of $-\arg T_{01}$, up to the limit of inequalities (7.6) and (7.7) charted in Fig. 7.12, will make the response for T_{01} nonmonotonic, thus contradicting (7.4).

7.2.5 Modifications

Beyond the frequency band of operation, the ratio V_{s2}/V_{s1} in the summing point can be appreciably magnified by making the threshold of saturation frequency-dependent in either the main or the auxiliary channel. This is accomplished, for example, in the system of Fig. 7.15 by introduction of a lowpass link Q at the output of the main channel.

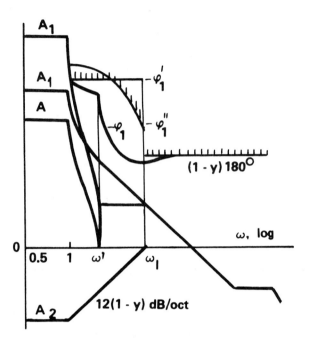

FIG. 7.14

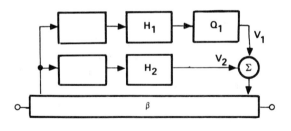

FIG. 7.15

In Fig. 7.16, let the gain of linear compensators Q_1 and Q_2 decrease with frequency within the frequency band $[1, \omega_l]$ (the compensators may be those leveling off the excessive loop gain while realizing the concave profile of the Bode diagram). In this case, saturation can occur in the penultimate stages. If $|Q_1|$ is small enough, a fairly high ratio V_{s2}/V_{s1} can be attained. It was shown by Beliavtsev and Zhukov [11] that this benefit increases with the increase of v_{s2}/v_{s1} (and $y_2 - y$) in the frequency band of operation.

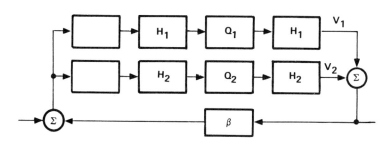

FIG. 7.16

In Figs. 7.17 and 7.18, the summing point is transferred into the feedback path. Because the signal reaching the summing point from the main channel is attenuated by the passive circuit, a fairly high ratio v_{S2}/v_{S1} is easily attained. However, the local feedback introduced around the first stage lowers the loop gain around the last stage of the main channel. For this reason, such block diagrams can only be recommended if H_2 comprises a dead zone, and does not operate when the signal is under the saturation threshold of the main channel (see the study in both the next section and the next chapter), or if the power stage of the main channel is followed by a lowpass filter (see next chapter).

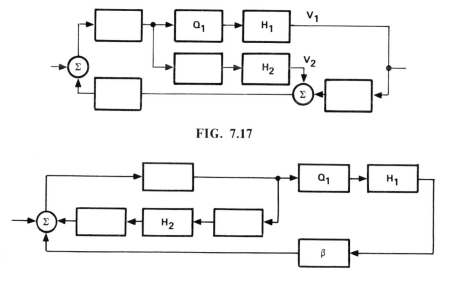

FIG. 7.17

FIG. 7.18

Combining a NDC in the local feedback with an auxiliary parallel channel might be advantageous as compared with the system having only a NDC or a parallel auxiliary channel. Consider, for example, the feedback amplifier shown in Fig. 7.19.

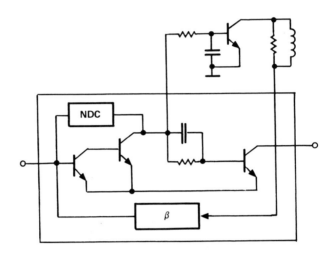

FIG. 7.19

Three low-power high-frequency transistors Q_1, Q_2, and Q_3, arrayed closely together in a hybrid circuit, constitute an auxiliary feedback channel with the best achievable asymptote. The main channel output transistor Q_4 is of high power. Due to its larger physical dimensions, the asymptote of the main channel, comprising transistors Q_1, Q_2, and Q_4, is worse than the asymptote of the auxiliary channel.

In the channel thusly composed, the high-frequency asymptote is better than in the main channel taken alone. Then, the stability in the linear state of operation can be guaranteed with increased feedback in the working band. The stability conditions for the large-amplitude signals at high frequencies (let us say, $\omega > 2$) is guaranteed with rather low power of the auxiliary channel, due to passive correctors introduced in the interstage circuits.

At $\omega < 2$, the oscillation with large amplitudes is eliminated by means of a NDC in the local feedback path. In this way, the ratio v_{s2}/v_{s1} need not be large.

7.2.6 Numerical Examples

Consider the design of a feedback amplifier containing three stages of amplification in each of the two channels having equal high-frequency asymptotes. Feedback of 37 dB is available in each of the separate channels with a phase stability margin of $\pi/6$ and an amplitude stability margin of 10 dB.

In the two-channel system, with $x_1 = x_2 = 16$ dB, the frequency $\omega_l = 4.1$, and $20 \log|T_{01}(j\omega_l)| = 20 \log|T_{02}(j\omega_l)| = 4$ dB. At $\omega \geq \omega_l$, $|T_{01}|$ and $|T_{02}|$ coincide. At $\omega = \omega_l$, their frequency responses depend on the ratio v_{s2}/v_{s1}. Consider two examples:

1. Let $v_{s2}/v_{s1} = 0.7$. Using (7.16), $-\phi_2 = 80°$, and (7.3) yields $\omega_r = \omega_l 80°/150° = 0.53 \omega_l = 2.2$. The slope of the Bode plot for T_{02} is $12 \cdot 80°/180° = 0.53$ dB/octave. The resulting Bode plot for T_{02} is depicted in Fig. 7.20.

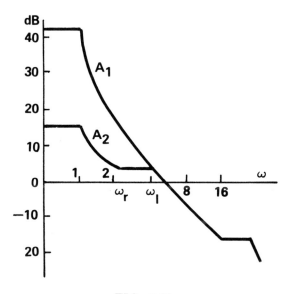

FIG. 7.20

From (7.14), or from the charts of Fig. 7.12, we can see that at $\omega \in [1, \omega_l]$ the phase shift $-\phi_1 = 190°$. Therefore, in the region $\omega \leq \omega_l$, T_{01} can be realized as a Bode cut-off with the main slope of -12.7 dB (related to the phase shift $-190°$), with 42.4 dB of $|T_{01}|$ in the band of operation. The benefit in feedback of the two-channel system over that of the single-channel system constitutes 5.4 dB.

In this example, no steps were implemented on the Bode plot for T_{01} and T_{02} near the frequency ω_l, since, due to small $-\phi_2$, the $\arg(T_{01} + T_{02})$ satisfies the phase stability margin.

2. Let us choose $(y_2 - y)180° = 8.5°$, i.e., $y_2 180° = 38.5°$, and construct the Bode diagram for T_{02} using the relations of Sec. 7.2.1 (assuming $|T_{02}|$ to be constant in the band of operation).

From (7.10), $-\phi_{(kl)}(\omega_k) = 240°$, and $-\phi_{(kl)}(1) = 317°$. From (7.8), or from Fig. 7.12, $-\phi_{sens} = 290°$.

The Bode plot of T_{01} satisfying these limitations was composed from the Bode constant-slope segments (see Sec. 2.3.2). The related Nyquist diagram for $T_{01} + T_{02}$ is charted in Fig. 7.21. The benefit in feedback over that of the single-channel system constitutes 17 dB.

Provided $v_{S2}/v_{S1} \gg 1$, the calculated AAPC $T_1 + T_2$ are shown in Fig. 7.21. The minimal value for the ratio v_{S2}/v_{S1} is 6.7 in accordance with (7.14) or Fig. 7.12. This value may be reduced if, as mentioned above, y_2 is slightly increased near the frequency ω_k, and if the saturation in the channels is made frequency-dependent. This was accomplished in the experimental device described in the next section.

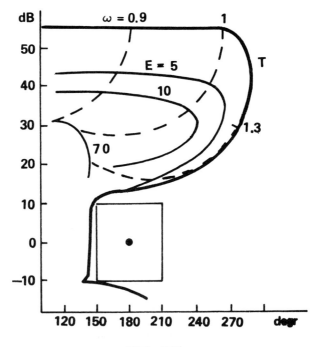

FIG. 7.21

7.2.7 Experiment

Figure 7.22 shows the experimental two-channel feedback amplifier [13] whose parameters were theoretically calculated in *Example 2* of the previous section.

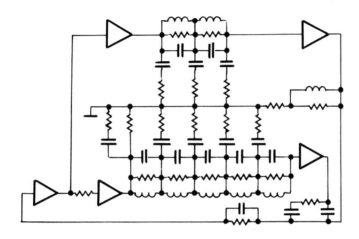

FIG. 7.22

Each channel contains three amplification stages using transistors with $f_T = 120$ MHz, the first being the common stage. The measured frequency responses for T_{01} and T_{02} charted in Fig. 7.23 were formed by interstage correctors in accordance with the block diagram of Fig. 7.16. The measured Nyquist diagram and the AAPC for $T_1 + T_2$ are plotted in Fig. 7.24.

At $\omega = 1$, the phase shift $-\phi_l$ achieves 285°. In order for the related AAPC to avoid the stability margin, the ratio v_{S2}/v_{S1}, as follows from (7.14) or Fig. 7.12, must exceed 3.9. This was accomplished.

At frequencies $\omega > 1.3$, the ultimate stage cannot be saturated because the attenuation of the interstage correctors becomes rather large. Therefore, the condition $v_{S2}/v_{S1} \geq 6.7$ should be applied to the saturation produced in the subsequent stages. This condition was met in the experimental device.

The first-stage nonlinearity does not affect the stability conditions because of its sufficient output power capability.

The improved amplifier linearity (with the second-harmonic attenuation of 56 dB) affirms the value of the return ratio measured directly.

The system seemed to be globally stable. The minimal E'_a throughout the band of operation (at $\omega = 1$) was 0.8, which is quite satisfactory.

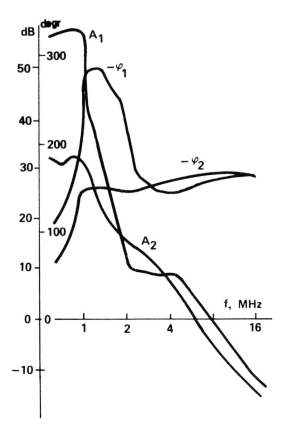

FIG. 7.23

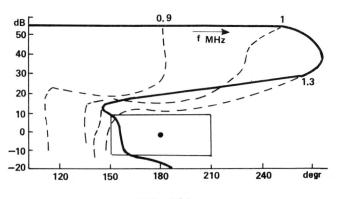

FIG. 7.24

7.3 TWO PARALLEL CHANNELS WITH SATURATION AND DEAD ZONE

7.3.1 Frequency Responses of the Linear Part

A dead zone in the auxiliary channel reduces the power consumption (and the sensitivity to the first-channel gain variations). If the zone equals the saturation threshold of the main channel, the auxiliary channel will be off during the normal state of operation, and only working with high efficiency when the main channel is overloaded.

The analysis and design of such a system is similar to the procedure already described for the system without dead zone. The Bode diagrams for T_{02} at all frequencies and for T_{01} at $\omega \geq \omega_l$, will not differ from those found previously (shown in Fig. 7.9), having the same amplitude stability margin in each channel of $x + 6\,\text{dB}$ (x_1 cannot be reduced for the system with dead zone because, when the signal amplitude increases, $|T_1 + T_2|$ rises at higher frequencies up to $\sim 2|T_1|$).

The upper limit for $-\phi$ at the frequencies $\omega \leq \omega_l$ is found to be as previously, through considering the AAPC for

$$T = T_{01} H_1(E) + T_{02} H_2(E) \tag{7.18}$$

As mentioned, over the working frequency band, the width of the dead zone equals the threshold of saturation of the main channel. Beyond the working band, due to frequency dependence of the channel gains, the width of the dead zone relative to the main channel saturation threshold (considered to be 1) is

$$e_d(\omega) = \left| \frac{\mu_2(j)}{\mu_1(j)} \frac{\mu_1(j\omega)}{\mu_2(j\omega)} \right| = \left| \frac{T_{02}(j)}{T_{01}(j)} \frac{T_{01}(j\omega)}{T_{02}(j\omega)} \right| < 1 \tag{7.19}$$

For $E < e_{s2}$, the auxiliary channel DF is characterized by the expression (3.26) for the dead zone.

We will examine piece by piece the AAPC for $T_1 + T_2$ shown in Fig. 7.25 using the simplified formulas given in Sec. 3.6.

1. On the first piece characterized by $E \epsilon [e_d, 1]$ and

$$T = T_{01} + T_{02}\,[1 - 1.27 e_d/E + 0.27 e_d^4/E^4]$$

only the coefficient at T_{02} changes with E. The piece of the AAPC is, therefore, a segment of a straight line parallel to the vector T_{02}.

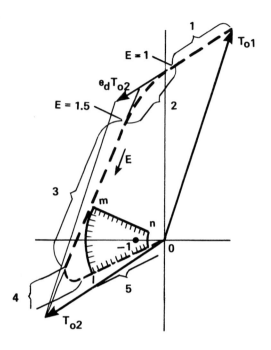

FIG. 7.25

Beyond the first piece, for $E \in [1, e_{S2}]$, T is given by using (3.26) and (3.27) as

$$T = T_{01}[1.27/E - 0.27/E^4] + T_{02}[1 - 1.27e_d/E + 0.27e_d^4/E^4] \qquad (7.20)$$

2. The second piece, for $E \in [1, 1.5]$ is comparatively short. At this piece, AAPC T determined by (7.20) is curvilinear. It can be seen from Fig. 7.25 that neither the first nor the second pieces of the AAPC are critical in determining T_{01}.

3. On the third piece, where $E \in [1.5, e_{S2}]$, the expression (7.20) reduces to

$$T \cong T_{01}\,1.27/E + T_{02} - 1.27\,T_{02}\,e_d/E = T_{02} + (T_{01} - e_d T_{02})1.27/E \qquad (7.21)$$

This piece of the AAPC T presents a segment of a straight line aimed at the end of the vector T_{02}.

Note that in the working band, where $e_d = 1$, the first piece of the AAPC is absent, and, therefore, the second and third pieces lie on the straight line passing through the ends of the vectors T_{01} and T_{02}.

Observations that $|T_{02}| < |T_{01}|$, that the amplitude E is large in the neighborhood of the arc kl, or that e_d is small, permit us to neglect the term $e_d T_{02}$ in (7.21) in the vicinity of the stability margin boundary. In other words, in this vicinity, the AAPC can be approximated by the side of the parallelogram shown in Fig. 7.25, and $\phi_{(kl)}$ may be determined as described in Sec. 7.2.2.

4. On the fourth piece, which is a rather short curvilinear section, $E \in [e_d, 1.5 e_d]$, T is determined through (7.18), (3.26), and (3.27), and the AAPC does not enter the stability margin if it does not do so at other sections. Hence, this section is not critical in determining the acceptable responses for T_{01} and T_{02}, thus, it does not deserve detailed analysis.

5. On the fifth piece, where $E \geq 1.5 e_{s2}$, and

$$T = 1.27 T_{01}/E + 1.27(e_{s2} - e_d) T_{02}/E = 1.27 [T_{01} + (e_{s2} - e_d) T_{02}]/E \quad (7.22)$$

the AAPC goes to the origin along a straight line. Expression (7.22) differs from expression (7.12) by only the multiplier $(1 - e_d/e_{s2})$ at T_{02}. Consequently, the previously obtained relationships (7.13) to (7.15) are applicable after replacing v_{s2}/v_{s1} by $(1 - e_d/e_{s2}) v_{s2}/v_{s1}$.

We conclude that the difference in determining T_{01} for the system with dead zone with respect to the system without the dead zone consists of increasing the lower boundary for the ratio v_{s2}/v_{s1} and, certainly, replacing the condition (7.4) by

$$|T_{01}| \geq 10^{x/20}$$

7.3.2 Jump-Resonance

An estimation of E'_a with (4.15) and (4.16) shows that E'_a belongs to the third piece of the AAPC T.

In the working frequency band (where E'_a should be delimited from below), $e_d = 1$, and from (7.21),

$$T = T_{02} + (T_{01} - T_{02}) 1.27/E$$

Exploiting considerations like those used in the analysis of the system without dead zone in Sec. 7.2.3, the system may be transformed into a single-loop system with the return ratio:

$$\frac{T_{01} - T_{02}}{T_{02} + 1} \cong \frac{T_{01}}{T_{02}}$$

From the diagram of Fig. 4.22, since $|T_{02}/T_{01}|$ is large, E' is fully determined by $|\phi_1 - \phi_2|$. For, let us say, $E'_a = 0.75$, there must be $|\phi_1 - \phi_2| = 38°$, which imposes an additional limitation on ϕ_1 in the working band.

7.3.3 Nonlinear Dynamic Compensator

As mentioned in Sec. 7.2.5, the auxiliary parallel channel can be equivalently transformed into a local loop. As a result, the content of the foregoing sections becomes just another approach to the synthesis of a system with NDC. The NDC in the local loop may be preferable because of less stringent requirements on the tolerances and accuracy of realization of the transfer coefficients of the linear links (recall *Problem 2* from Sec. 1.2.4).

7.3.4 Experiment

The experimental feedback amplifier [14] diagrammed in Fig. 7.26 is very similar to that of Fig. 7.22, except for the output stage of the auxiliary channel. The output stage is composed of two complimentary transistors, biased to form a dead zone on the amplitude transfer characteristic. The stability margins have been selected as $x_2 = x_3 = x = 10\,\text{dB}$ and $y_2 180° = 48°$. According to (7.14), or by using the diagram of Fig. 7.12, achievement of the maximum available feedback requires the ratio $v_{s2}/v_{s1} \geq 3.1$. Taking into account the multiplier $(1 - e_d/e_{s2})$, this ratio has to be somewhat larger; in the experiment, it equals 4.

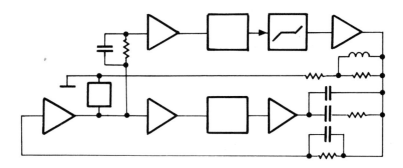

FIG. 7.26

The Bode diagram for T_{01} was calculated as described in Sec. 7.2.4. The selected parameters $\omega' = 1.6$, and $-B_{3m} = 180°$ result in $20 \log q = 30$ dB. At $\omega = 1$, the difference in the channel's phase shifts is $|\phi_1 - \phi_2| = 38°$, so the jump-resonance characteristic $E'_a = 0.75$. The benefit in the feedback over the single-loop system is 13 dB.

The theoretically defined frequency responses were realized with high accuracy. Figure 7.27 presents the experimentally obtained Nyquist diagram for the return ratio $T_{01} + T_{02}$ of the total amplifier and the AAPC T. As can be judged from the experiments, the system is globally stable. The mode and values of the jumps of the jump-resonance were in agreement with the theory.

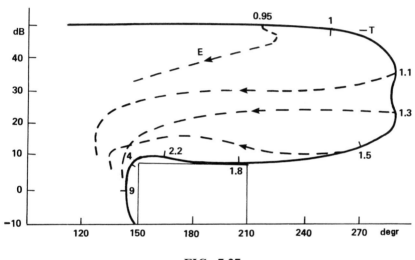

FIG. 7.27

Chapter 8
Nonlinear Multiloop System: Absolute Stability Approach

8.1 SYSTEM REDUCIBLE TO SINGLE-CHANNEL

8.1.1 Block Diagrams

The systems under study are shown in Figs. 8.1 and 8.2. The plant contains the nonlinear link $v(e)$ and the linear link Q. The preamplifier T_0/QG and the link G in the NDC are linear. The return ratio for the plant, which was measured while substituting the link v by a unitary link, is T_0.

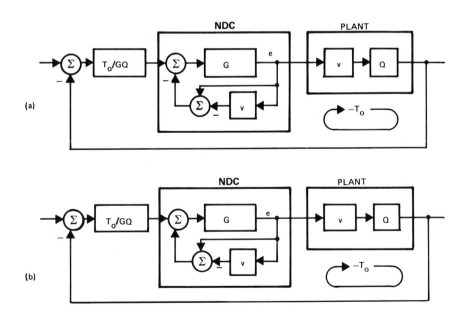

FIG. 8.1

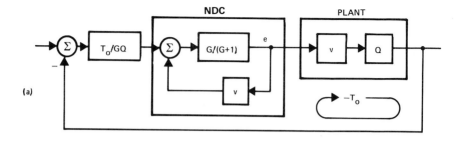

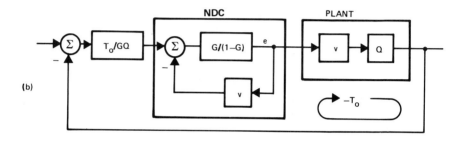

FIG. 8.2

We assume that the transmission functions G and T_0 are noninverting, i.e., in a lowpass system at $\omega = 0$, $\mathrm{Re}\, G = G > 0$ and $\mathrm{Re}\, T_0 = T_0 > 0$. In the bandpass system, these equalities hold at some "center" frequency within the passband.

The systems in Fig. 8.1(a,b) can be transformed correspondingly into those shown in Fig. 8.2(a,b). Further, these systems can be converted into the systems shown in Fig. 8.3(a,b), each containing only one nonlinear link $-v$. The return ratio measured for the cross section at the input of the link $-v$ is T_e. Thus, global stability of the system with the NDC is evidently achieved if T_e for the equivalent single-loop system satisfies the Popov criterion.

The following relations hold:

For the system of Fig. 8.3(a):

$$T_e = \frac{T_0 - G}{1 + G} \tag{8.1}$$

From (8.1), for the system of Figs. 8.1(a), 8.2(a), and 8.3(a):

$$G = \frac{T_0 - T_e}{1 + T_e} \tag{8.2}$$

Nonlinear Multiloop System — AS Approach

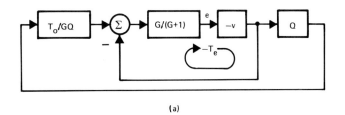

(a)

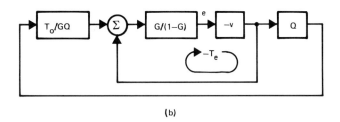

(b)

FIG. 8.3

$$\frac{G}{1+G} = \frac{T - T_e}{1 + T_0} \tag{8.3}$$

$$1 + T_0 = (1 + G)(1 + T_e) \tag{8.4}$$

Similarly, for the systems of Figs. 8.1(b), 8.2(b), and 8.3(b):

$$T_e = \frac{T_0 + G}{1 - G} \tag{8.5}$$

$$G = \frac{T_e - T_0}{1 + T_e} \tag{8.6}$$

$$\frac{G}{1 - G} = \frac{T_e - T_0}{1 + T_e} \tag{8.7}$$

$$1 + T_0 = (1 - G)(1 + T_e) \tag{8.8}$$

In the following sections, we proceed with the analysis of generic systems of Figs. 8.1 and 8.2, comprising two identical nonlinear links v. We shall keep in mind, however, that in practical applications a low-power link v', $v'(ae) = av(e)$, $a \ll 1$, can be employed instead of the link v in the NDC after equivalently transforming the systems as indicated in Fig. 8.4(a,b,c,d), respectively.

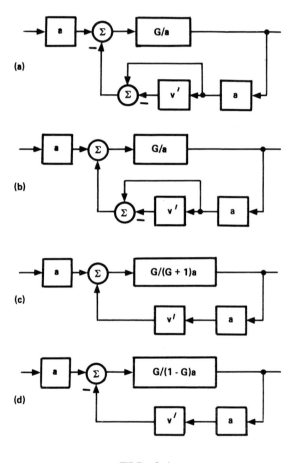

FIG. 8.4

8.1.2 Integral Constraint and Stability Margin

It would be realistic to suppose that T_0 and G decay at higher frequencies at least as ω^{-2}. Therefore, from (2.11),

$$\int_0^\infty \log|1 + T_e|\, d\omega = 0$$

$$\int_0^\infty \log|1 + G|\, d\omega = 0$$

Then, recalling (8.4), we have

$$\int_0^\infty \log|1 + T_0|\, d\omega = 0 \tag{8.9}$$

i.e., the area of negative feedback (when $|1 + T_0| > 1$) equals the area of positive feedback (when $|1 + T_0| < 1$) in the system considered, obeying the same law as the single-loop system. Therefore, in order to maximize the negative feedback in the band [0,1], the feedback $|T_0 + 1|$ must be minimized at higher frequencies.

Because the excessive positive feedback will lead to intolerably high sensitivity of the closed-loop system transmission, the feedback must be limited from below. In other words: (1) since the frequency hodograph for $T_0 + 1$ may not come too close to the origin, a certain stability margin must be introduced; (2) outside the band of operation, it is desirable that the locus $T_0 + 1$ approaches the boundary of this stability margin. The trial and error procedure executed in the examples to be presented in Sec. 8.2 aims to meet these two requirements.

There are several possible and fairly equivalent ways to define the boundary of the stability margin. The one chosen here has the advantage of analytical simplicity and restricts the permissible positive feedback by

$$20 \log|T_0 + 1| > p \tag{8.10}$$

where p in dB may be viewed as the admissible increase of the closed-loop system transmittance (and of its sensitivity) caused by the positive feedback. The resulting circular margin surrounding the critical point on the T_0-plane is compared to the phase and amplitude Bode stability margins in Fig. 8.5. The common value of the phase stability margin of 30° corresponds to $\min|T_0 + 1| = 0.5$, i.e., to $p = 6\,\text{dB}$. Clearly, we need to check (8.10) at $\{\omega: |T_0| \cong 1\}$ only.

8.1.3 Plant with Saturation

In order to clarify the practical aspects of the theory, we will treat the link $v(e)$ below as a saturation element $\{v = E: |e| < 1\}$, $\{v = 1: |e| > 1\}$, and the link composed of $-v$ and the unitary link in parallel, as a dead-zone element.

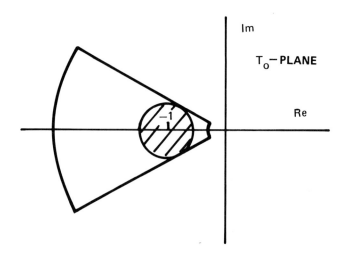

FIG. 8.5*

For $|e|<1$, the transfer function of the NDC is G, the return ratio for the plant is T_0, and the effect of small plant parameter variations is suppressed $|T_0+1|$ times. Then, maximization of $|T_0+1|$ is required.

Generally, the above specification of the nonlinear link is unnecessary. The systems to be examined are globally stable if they contain any nonlinear links $v(e)$ with $0<v/e<1$, or a link with "active hysterisis" [75].

8.2 DEAD-ZONE ELEMENT IN THE NDC FEEDBACK PATH

When the link v represents saturation in the plant, the NDC in the system of Fig. 8.1(a) is composed of a noninverting linear link looped by an inverting dead-zone link. We can see from (8.2) that for the feedback system to be realizable in a lowpass configuration having $|T_e| \gg 1$ and $|T_0| \gg 1$ across the entire operational band, the inequality $|T_0| > |T_e|$ should be fulfilled at zero frequency, where $T_e \gg 1$, $T_0 \gg 1$, and $G > 0$.

As mentioned above, the system is globally stable if the function $T_e(j\omega)$ satisfies the Popov criterion. In order to meet this condition while simultaneously maximizing the modulus of expression (8.4), we assume $T_e(j\omega) = T_B(j\omega)$, where T_B represents the optimal Bode cut-off with conventional stability margins of 30° and 10 dB.

---

*Figure 8.5 and other pictures in this chapter are reprinted from the author's papers [116, 117, 121] with permission of the publisher Taylor and Francis, Ltd.

Because T_e has now been determined, the frequency response of G does not influence the global stability conditions, which are already met. In order to maximize $|T_0+1| \cong |T_0|$ across the band [0,1], the value of $|G|$ ought to be as large as possible across this band. The frequency response $G(j\omega)$ to be constructed must satisfy the following realizability conditions:

1. It follows from considerations similar to those in Sec. 2.4 that $G(s)$ ideally should be a minimum phase function.
2. $G(j\omega)$ must be shaped in a manner permitting the Bode diagram of T_0 to intersect the 0 dB level to the left of the frequency ω_b. The feedback would certainly be larger if the crossover frequency is at its limit, i.e., if $|T_0(j\omega_b)|=1$.
3. The inequality (8.10) must hold.

The function $T_B(s)$ and $G(s)$ are transcendental and their frequency responses are of a very peculiar shape. They can be sufficiently approximated by rational functions of high order: 10 to 20. Although such an approximation and the hardware or software implementation represent little difficulty, finding the optimal rational function $G(s)$ is a rather complicated procedure. Instead, we employ an iterative procedure to find physically realizable transcendental functions, starting with T_B and some guessed response for $|G|$. After completing several iterations, the physically realizable response for G is found as a linear combination of several known, physically realizable responses. Finally, the desired responses for the linear links of the loop are approximated by realizable rational functions.

The procedure utilizing the Bode piecewise-linear approximation for computing the phase responses is illustrated below.

Example 1. Absolutely stable (AS) lowpass system. Let $\omega_b = 8$, the phase stability margin equals 30° and the amplitude stability margin equals 10 dB. The Bode plots for T_B and for T_B+1 are shown in Fig. 8.6.

The attained feedback in the AS system without the NDC is then 40 dB. In order to ensure AGS in the system with the NDC we assume $T_e = T_B$. Now, the appropriate frequency response for G must be found by a trial and error procedure.

Trial 1. As a point of departure, it is reasonable to assume $G = T_B$. In this case, according to (8.4), the feedback in dB, $20 \log |T_0+1| = 40 \log |T_B+1|$, doubles the ultimate attainable feedback of the single-loop system. Since $|T_0(j\omega_b)+1| = |T_B(j\omega_b)+1|^2 = 1/4$, then, approximately, $|T_0(j\omega_b)| = 1$, as required. However, due to excessive positive feedback in the frequency region near ω_b, the condition (8.10) requiring $p = 6$ dB is not satisfied. Figure 8.7(a) displays the Nyquist diagram for T_B+1 and T_0+1 (not to scale); the diagrams in the vicinity of the origin are plotted in Fig. 8.7(b). Because T_B is the return ratio of a stable closed-loop system, the function T_B+1 has no zeros in the

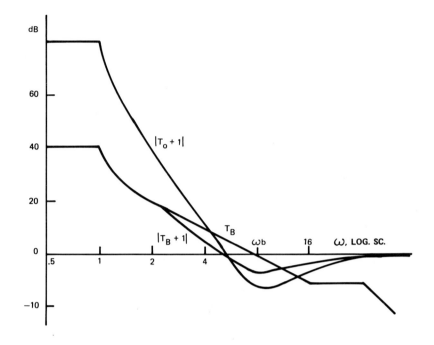

FIG. 8.6

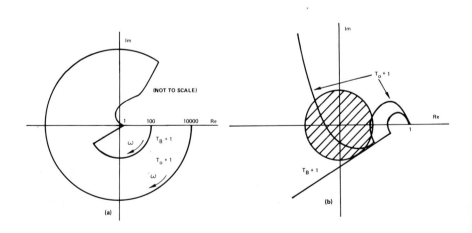

FIG. 8.7

right half-plane of the operational variable s, and obviously neither does $T_0 + 1 = (T_B + 1)^2$. Therefore, neither the locus of $T_B + 1$, nor the locus of $T_0 + 1$, encircle the origin. Further, because $\arg(T_0 + 1) = 2\arg(T_B + 1)$, the system is Nyquist-stable.

As Fig. 8.7(b) shows, the locus of $T_0 + 1$ enters the forbidden region of the stability margin. The reason is that both $\min|G+1|$ and $\min|T_e+1|$ occur at the same frequency near ω_b, thus, unfortunately, resulting in the pronounced minimum of $|T_0+1|$.

Trial 2. We assume $|G|$ lies on $|T_B|$ at $\omega \geq 16$, but at lower frequencies $|G| < |T_B|$, as Fig. 8.8 shows. Now the frequencies are different, at which $|G| = 1$ and $|T_B| = 1$, and, in addition, $|\arg G|$ is smaller in the vicinity of ω_b than before. As a result, the inequality (8.10) holds as well as the other realizability conditions. The attained benefit in feedback is 25 dB.

We can see from Fig. 8.8 that the area of positive feedback still can be increased, so we proceed further.

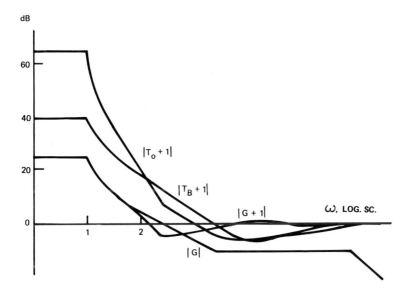

FIG. 8.8

Trial 3. We assume $|G| = |T_B|$ in the band $[0,1]$ and $G(j\omega)$ having sharper cut-off (which corresponds to the Nyquist-stable local loop in the NDC), as shown in Fig. 8.9. The feedback in dB doubles the attainable feedback in the conventional system without NDC, and all the realizability conditions

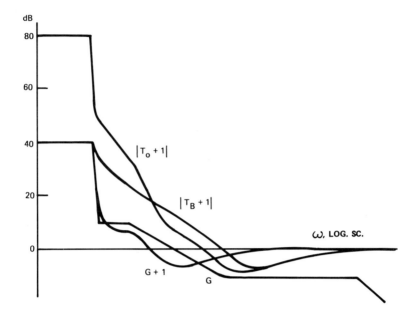

FIG. 8.9

are satisfied. The benefit in feedback is significant: 40 dB. A further substantial increase in feedback can only be achieved at the price of a nonmonotonic Bode diagram for T_0, which intersects the 0 dB level several times in the region $(1, \omega_b)$; this is usually impractical.

In this example, ω_b and $|T_B|$ were assumed to be fairly large. If ω_b and $|T_B|$ are smaller, the benefit in feedback is less. The benefit is approximately given by $20 \log |T_B| - 10$ dB.

The above example permits us to make an important conclusion: if the Popov criterion is used as the basis for design and the allowable feedback in a single-loop system is fairly large, then the allowable feedback in a system with the NDC is significantly greater than that in a single-loop system without the NDC.

Example 2. Lowpass Process-Stable System. As stated in Chapter 4, the inequality (4.3):

$$\text{Re}\,(T_e + 1) > 0$$

is the necessary and sufficient condition for the absolute stability of the process (APS) at the output of the single-loop system of Fig. 8.3. The fulfillment of this condition makes the system totally free from all forms of forced oscillation: jump-resonance, subharmonics, *et cetera*, and practically eliminates wind-up. However, this inequality constrains arg T_e, thus preventing the attainment of large feedback. That is why APS is rarely preserved in practical control systems.

By applying the NDC, the process instability can be entirely eliminated without reducing the feedback. Figures 8.10(a) and 8.11 show the diagrams for $G = T_B$ and T_e satisfying the inequality. The attained feedback $|T_0 + 1|$ of 50 dB exceeds that of the single-loop APS system and even of the single-loop AS system (40 dB).

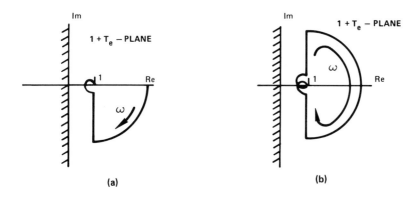

FIG. 8.10

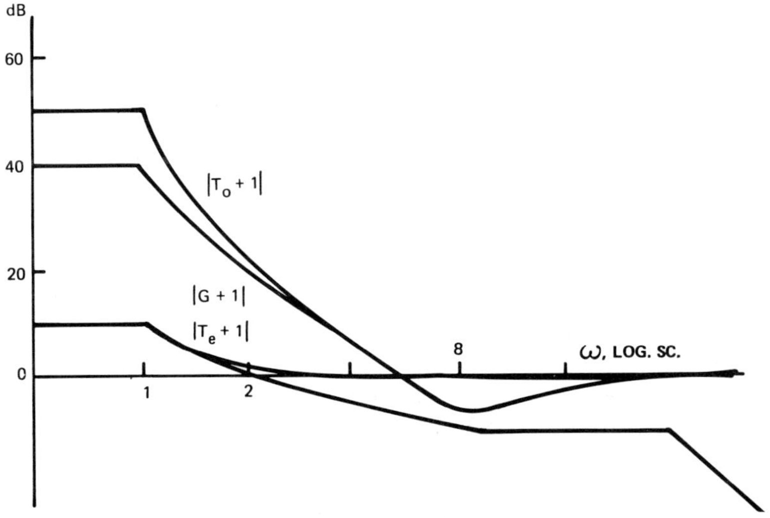

FIG. 8.11

This example demonstrates that in the lowpass APS system with NDC, the utmost available feedback is much more than that in the APS single-loop system, and may be even greater than the feedback in the AS (not APS) single-loop system.

Example 3. Bandpass System. As indicated in Sec. 2.4.1, a bandpass transform of the lowpass Bode optimal cut-off does not change the operational bandwidth. When, however, the phase stability margin in the lowpass prototype is less than $\pi/2$ (which is common), the transformed system with a single nonlinear nondynamic link does not satisfy the Popov criterion, as shown in Chapter 3. To meet this criterion, the steepness of the positive slope of the Bode diagram (at lower frequencies) must be reduced to approximately 6 dB/octave, thus cutting down the feedback in the bandpass system. The smaller the relative bandwidth, the greater is the defect in feedback.

The application of the NDC resolves this problem. For example, using the locus $T_e + 1$, shown in Fig. 8.10(b), which is the bandpass transform of the plot shown in Fig. 8.10(a), and using the Bode diagrams, which are the bandpass transforms of those shown in Fig. 8.10, we can design an APS bandpass system with the feedback (50 dB) greater than that attainable in the AS lowpass single-loop prototype (40 dB).

This study reveals that application of the NDC in an APS bandpass system increases the attainable feedback over the value achievable in a lowpass AS single-loop prototype.

8.3 POSITIVE AND NEGATIVE FEEDBACK IN THE NDC

We found in the previous section that in order for the feedback around the plant to be large, $|G|$ should be large, such that $G/(G+1) \cong 1$ across the band of operation. Therefore, the local loop transmission function in the NDC of Fig. 8.2(a) approaches 1 in the operational band, thus magnifying the sensitivity of the NDC gain in the linear state of operation. Hence, the transmission function of link $G/(G+1)$ must be realized with high precision. The most practical way of doing this is to apply a large internal feedback, i.e., to realize the link $G/(G+1)$ as the link G looped with unitary inverting feedback. This results, however, in the system already studied in Fig. 8.1(a).

Loosely speaking, the action of the NDC in Figs. 8.2(a) and 8.1(b) is in changing the positive feedback in the local loop by varying the DF of the nonlinear link v. As such, the two NDCs suffer from high gain sensitivity to variations in the transmission functions of the links in the local loop. Conversely, increasing the local negative feedback at lower frequencies in the NDC of Fig. 8.1(a) when the signal amplitude grows large or reducing the negative feedback at higher frequencies in Fig. 8.2(b) will ensure global stability without increasing the sensitivity.

Consider the system of Fig. 8.2(b) in more detail. In order for local loop transmission to be inverting, the link $G/(1-G)$ must be noninverting, i.e., at $\omega = 0$ $|G| \ll 1$ and $|T_e| > |T_0|$.

The goal is to maximize $|T_0|$ over the band of operation, with ω_b prescribed. Then, the frequency response for T_0 must be Nyquist-stable, with the Bode diagram for $T_0 + 1$ like those shown in Figs. 8.7, 8.8, and 8.9. For the system to be globally stable, let $T_e = T_{B1}$, where T_{B1} is the Bode optimal cut-off with the crossover frequency ω_{b1}. In order to attain $|T_e| > |T_0| = T_B$ over the operational band, ω_{b1} must certainly be greater than ω_b, in practice, 4 to 100 times. After having determined the frequency responses for T_e and T_0, the response for the link $G/(1-G)$ is found from (8.7).

Over the band $\omega \in [4\omega_b, \omega_{b1}]$, it is required $|T_0| \ll 1$ and $|T_0| \ll |T_e|$. Then, from (8.7), $|G/(1-G)| \cong |T_e| = |T_{B1}|$. The link $G/(1-G)$ is, therefore, required to provide considerable gain at the frequencies up to $\omega_{b1} \gg \omega_b$. Such a requirement restricts the application of the block diagram to only those plants which include a lowpass with a cut-off frequency that is much lower than the cut-off frequencies of the rest of the links of the system. In this kind of system, the frequency ω_b is typically limited by the noise contributed by the sensor and the amplifier. The theory and the experiment concerning this system will be described in Sec. 8.5.

With v being a saturation link, the block diagram of Fig. 8.2(b) can be modified into the equivalent diagram of Fig. 8.12. In a practical realization of this system with appropriate scaling in the signal level (similar to that shown in Fig. 8.4), the amplifier in the NDC may also conveniently perform the role of a limiter.

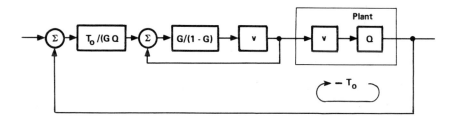

FIG. 8.12

8.4 DF *versus* AS TECHNIQUES

A synthesis procedure based on AS methods is only correct if applied to a system satisfying A. Lurie's assumptions, i.e., with v strictly nondynamic and all other links linear. However, the deviation of a practical system from this idealization might be substantial. To assert the AGS and to refine such system parameters, the methods of functional analysis or DF methods may be employed. As pointed out in Chapter 5, the DF stability analysis is fairly accurate when applied to a Nyquist-stable system having a filter-like loop response.

The DF analysis should, therefore, be used as a complement to AS methods. It will be employed in Sec. 8.5.4 to find the maximum admissible discrepancies between the characteristics of the two nonlinear links, in the plant and in the NDC. For this reason, we will examine the DF of the systems under study.

While E_c grows bigger, ω_c decreases (the subscript c, introduced in Sec. 5.2.2, marks the values related to the case of $|T|=1$). Thus, for the system to be globally stable, i.e., for the $|\arg T_c| < \pi$, the frequency range where the slope of the Bode diagram for the equivalent linearized system (see Sec. 5.2.7) is required to be small tends toward the lower frequencies. In other words, while increasing E_c, the lower-frequency part of the diagram must decrease faster than the higher-frequency part. The explanation of how this works in the systems of Figs. 8.1(a) and 8.2(b) follows.

In the system of Fig. 8.1(a), increasing E equally reduces the DF of the plant at all frequencies and simultaneously reduces the DF of the NDC, but for the NDC at lower frequencies only, as the result of rising local negative feedback.

If, for example, $E=2.5$, the DF of the saturation element v is 0.5, as seen from (3.25), and, therefore, the DF of the NDC decreases by $\sim |0.5G+1|$ times. Simultaneously, $|T|$ decreases by 2 times due to saturation in the plant, reducing $|T|$ by $|G+2|$ times. At higher frequencies, where $|G|$ is smaller, the decrease in $|T|$ is smaller as well. Therefore, the desired feature is realized of the Bode plot for T.

The decrease of $|T|$ at higher frequencies (6 dB in the above example) is undesirable because it makes the slope of the Bode diagram and the related phase lag larger than would be if the high-frequency loop gain were unchanged. The performance is better for the system of Fig. 8.2(b), where increasing E reduces the plant transmission over a very wide frequency band, while reducing the local negative feedback in the NDC at higher frequencies only. Hence, $|T|$ decreases at lower frequencies and remains unchanged at higher frequencies.

Comparing the NDC designed with the AS approach with that described in Chapter 6 with the DF method discloses that the NDC of Fig. 8.1 is a simplified version of the NDC of Fig. 5.13, without the upper branch. Its transmission DF, therefore, has the form $\alpha/(\omega+\gamma)$, with a smaller potential increment in the phase shift as compared with (5.6). However, the synthesis procedure based on the AS approach is not restricted by the conditions (5.9) and (5.16), therefore, the NDC is allowed to be intrinsically conditionally stable, which permits ψ_c to exceed π.

Generally speaking, even for DF methods, these limitations need not be satisfied. We may locally optimize the initial crude solution achieved under these limitations by varying the frequency responses of the linear links under the control of the sufficient stability criterion given in Sec. 5.2.7.

8.5 TWO-CHANNEL SYSTEM WITH A LOWPASS IN THE PLANT

8.5.1 Block Diagram

As noted in Sec. 8.3, the system shown in Fig. 8.2(b) can be employed to increase the available feedback around the plant, which contains a lowpass link at the output, and whose feedback is limited by the noise at the input of the actuator.

By shifting the second point of summing equivalently backward in the system of Fig. 8.2(b), we can transform it into a system with two parallel channels, each channel having a similar nonlinear link v (to be understood in the following as saturation elements), as shown in Fig. 8.13. Further, we

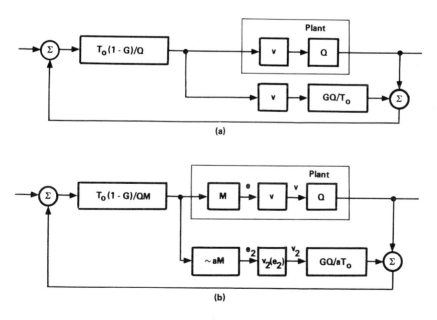

FIG. 8.13

approximate the system of Fig. 8.13(a) by that of Fig. 8.13(b), by including a linear link M into the plant and exploiting, as mentioned in connection with Fig. 8.4(d), a low-power subsidiary channel with nonlinear link v' having a transfer characteristic for the signal $e_2 = ae$, which is the same that v has for the signal e,

$$v_2(e_2) \cong av(ae) \tag{8.11}$$

where a is a small real constant.

Allowing $a = 10^{-2}$ causes the amplitudes of the nonlinear links outputs to differ by 10^2 and the powers to differ by 10^4. Assuming, further, that the responses of M and Q of the plant are known, and using the frequency responses for G and T_0 as those given in Sec. 8.2.2, we are in a position to determine the frequency responses of all the linear links in the practical system (with a low-power auxiliary channel) of Fig. 8.13(b).

It is also instructive to examine this system, redrawn in Fig. 8.14(a), as a two-channel system having the frequency responses of the type depicted

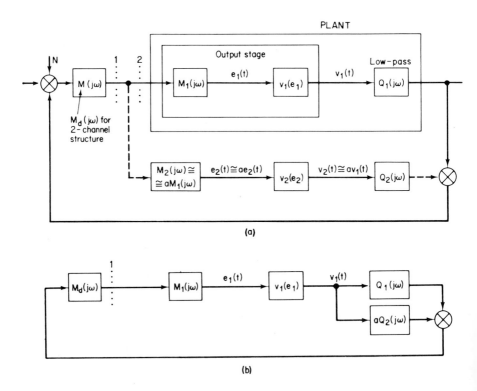

FIG. 8.14

in Fig. 8.15. The transfer functions of all the linear links are considered MP. In the two-channel structure, the preamplifier M_d replaces the amplifier M of the optimal single-loop system. The auxiliary channel is shown by the dashed lines. The system of Fig. 8.14(b) approximates the system of Fig. 8.14(a). This system contains a single nonlinear link, so its global stability can be studied by the scalar Popov criterion.

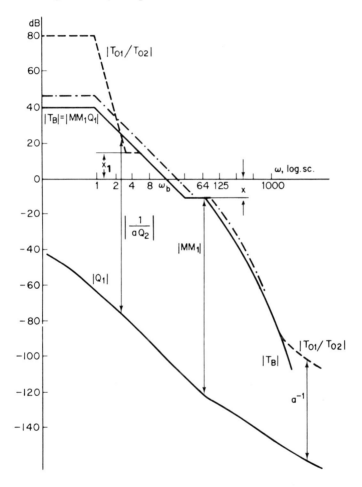

FIG. 8.15

Denote the individual channels return ratios as $T_{01} = M_d M Q$, $T_{02} = M_d M_2 Q_2 = M_{da} a M Q_2$. In the linear state of operation, the return ratio for the equivalent system of Fig. 8.14(b) (which can be measured at the cross section point 1) is $T_e = T_{01} + T_{02}$. For the T to be MP, the locus of $T_{01}(j\omega)/T_{02}(j\omega)$ must avoid the point $(-1, 0)$ with some safety margin (recall Sec. 2.5.2). With the function T_e being MP, the Bode phase-gain relations apply, and the Popov criterion is satisfied when $T_e = T_B$, where T_B, as before, is the optimal Bode cut-off.

The system transfer function is

$$\frac{T_{01}}{T_{01} + T_{02} + 1}$$

The sensitivity to MQ is

$$S_1 = \frac{T_{02} + 1}{T_{01} + T_{02} + 1}$$

and the sensitivity to $M_2 Q_2$ is

$$S_2 = \frac{-T_{02}}{T_{01} + T_{02} + 1}$$

It is clear, that in order to obtain small $|S_1|$, we must have $|T_{01}| \gg |T_{02} + 1|$. In this case,

$$S_1 \cong \frac{T_{02} + 1}{T_{01}} = \frac{T_{02}}{T_{01}} \left(1 + \frac{1}{T_{02}} \right)$$

As shown below, maximum of $\min(\omega \epsilon [0,1])\ |T_{01}/T_{02}|$ is limited. Consequently, in order to reduce $|S_1|$, we must have $|T_{02}| \gg 1$, and then

$$S_1^{-1} \cong -S_2^{-1} \cong T_{01}/T_{02}$$

Therefore, only $|T_{01}/T_{02}|$ has to be maximized over the operational band $\omega \leq 1$, and the responses that characterize the Nyquist-stable system (see Sec. 2.4.2) for T_{01}/T_{02} give a solution which is nearly optimal.

Figure 8.16 presents numerical examples for the frequency responses of $|T_{01}|$, $|T_{02}|$, and $|T_{01} + T_{02}|$, and the hodograph for T_{01}/T_{02}. Here, $\arg T_{01}(j\omega_b)/T_{02}(j\omega_b) = -\pi/3$, and since $T_{01}(j\omega_b) = T_{02}(j\omega_b)$, at $\omega = \omega_b$, $|T_{01} + T_{02}| \cong |T_{02}| \cong |T_{02}| \cdot |T_{01} + T_{02}| \cong |T_{01}|$ for $\omega < \omega_b$ and $|T_{01} + T_{02}| \cong |T_{02}|$

for $\omega \geq \omega_b$. It can be seen that in the band $[0, 1]$ $20 \log |T_{01}/T_{02}| - 20 \log |T_B| = 40$ dB, i.e., the sensitivity S_1 in the two-channel system is 40 dB better.

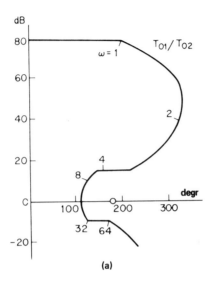

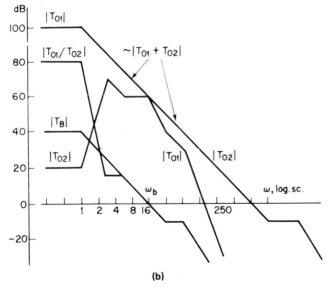

FIG. 8.16

8.5.2 Noise

We will now prove that in a two-channel system with frequency response T_{01}/T_{02}, as in the above, the noise amplitude at the input to the link v:

$$E_{Nd} = N \left| \frac{M_d}{1 + T_{02}} \frac{M_1}{1 + T_{01}/(T_{02} + 1)} \right|$$

at all frequencies is smaller than or equal to the noise amplitude in the single-loop version:

$$E_N = N \left| \frac{M M_1 T_{02}}{T_0 (T_B + 1)} \right|$$

Because $M_d = M T_{01}/T_B$ (recall $T_B = M M_1 Q_1$),

$$E_{Nd} = N \left| \frac{M M_1 T_{01}}{(T_{01} + T_{02} + 1) T} \right|$$

and

$$\frac{E_{Nd}}{E_N} = \left| \frac{T_{01}(T_B + 1)}{(T_{01} + T_{02} + 1) T_B} \right| = \frac{|(T_{01}/T_{02})(T_{01}/T_{02} + 1 + 1/T_{02})|}{|T_B/(T_B + 1)|} \quad (8.12)$$

Because $|T_{02}| \gg 1$ across a wide frequency band (see Fig. 8.16), we may neglect the term $1/T_{02}$ in this band. For $\omega < \omega_b/2$, the numerator and denominator in (8.12) approach 1. Hence, their ratio (8.12) approximates 1.

In proximity to ω_b, $T_{01}/T_{02} \cong T_B$, and expression (8.12) degenerates to ~ 1. (More precisely, it is less than 1 when the phase safety margin for T_{01}/T_{02} exceeds the phase stability margin for T_B, which can be accomplished.)

In the frequency band where $1 \gg |T_B| > Q_1/a$ (for $30 < \omega < 1000$ in Fig. 8.16), we can make $T_{01}/T_{02} = T_B$, so that (8.12) approximately equals 1.

At higher frequencies, $|T_{01}/T_{02}|$ cannot decrease as rapidly as $|T_B|$, since $|T_{01}/T_{02}| > |Q_1/a|$. However, as noted previously, the noise components in this frequency band are not significant. In addition, in this range, first, $|T_{02}|$ is small, and, second, expression (8.12) decreases due to the term $1/T_{02}$.

8.5.3 Other Points of View

1. The auxiliary channel in Figs. 8.13 and 8.14 can be considered as a nonlinear feedback path for the preamplifier M_d forming the NDC as shown in Figs. 8.2(b) and 8.17.

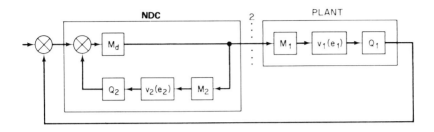

FIG. 8.17

In the wide frequency range where $|T_{02}| \gg 1$, the loop transmission function measured in the cross section 2 is

$$\frac{T_{01}}{T_{02}+1} \cong \frac{T_{01}}{T_{02}}$$

the Nyquist diagram for which is shown in Fig. 8.16. The system is Nyquist-stable. If the nonlinear compensator consisted of only linear elements, the system would be conditionally stable. However, we have proved that a properly designed system with two nonlinear elements is globally stable. Consequently, the NDC provides the unconditional (global) stability for the Nyquist-stable system.

2. According to the Bode-Nyquist multiple-loop stability criterion, if the system in Fig. 8.14 is stable in the linear mode of operation, the total number of clockwise and counterclockwise encirclements of the critical point must be equal to each other in the two Nyquist diagrams for T_{02} and $T_{01}/(T_{02}+1)$. The Nyquist diagram for $T_{02}/(T_{02}+1)$ does not encircle the critical point and, hence, neither does the diagram for T_{02}. Therefore, the system is stable while switching off the main channel (although perhaps conditionally stable); it is unstable without the auxiliary channel; and the two-channel system, as noted, is globally stable.

3. In the case of a digital link or a pulsed power stage in the main channel, a lowpass filter Q_1 smooths the shapes of the output signal. If the sampling frequency $\omega_s \gg \omega_b$, the theory presented above can be applied because in the band $\omega < \omega_b$ the output of the auxiliary channel can be neglected, and in the band $\omega > \omega_b$ the output of the auxiliary channel dominates in the sum of the two channels.

4. The structure of Fig. 8.14 can be simplified when it is possible and convenient to place an auxiliary sensor at the input of the link Q_1. The linear

circuit aQ_2 then provides a high-frequency bypass, reducing the attenuation of the composite link $aQ_2 + Q_1$ compared with that of Q_1, hence, reducing the noise.

5. With a sampling element in the main channel, the frequency ω_s must be of order $3\omega_b$ (or greater). This requirement is significantly less critical than the similar requirement for a single-channel system with comparable frequency responses, where the crossover frequency is much higher, being equal to the frequency in the two-channel structure at which $|T_{01} + T_{02}| = 1$.

8.5.4 The Allowable Discrepancy of Nonlinear Link Characteristics

In practice, (8.11) cannot be satisfied exactly. Hence, in order to show that the system is robust, the maximum allowable discrepancy of the two nonlinear link characteristics must be found. For this purpose, we will use the describing function (DF) analysis.

The discrepancy of the nonlinear link characteristics influences the performance of the systems of Fig. 8.18(a) and Fig. 8.18(b) in different ways.

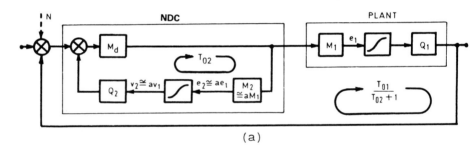

(a)

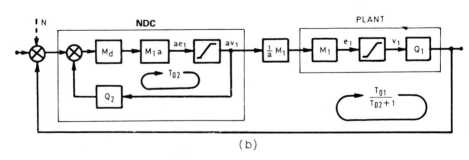

(b)

FIG. 8.18

Let us consider first the system of Fig. 8.18(b). In this system — of which the systems of Figs. 8.19, 8.20, and 8.21 are the models — the nonlinear links are connected in cascade. Therefore, if the threshold of saturation in the NDC is smaller than that of the following nonlinear link, the system can be considered to have a single (first) nonlinear link and, according to the Popov criterion, it is globally stable.

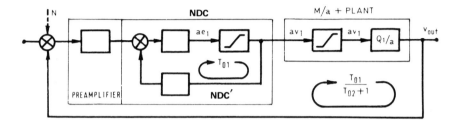

FIG. 8.19

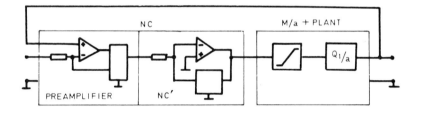

FIG. 8.20

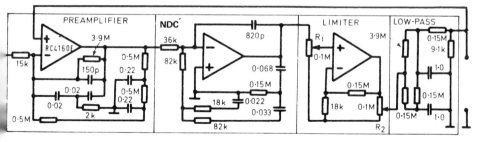

FIG. 8.21

Let H_1 and H_2 be the describing functions for the nonlinear links. If there is oscillation, the transmittance of the fundamental around the loop becomes 1, i.e.,

$$1 = -\frac{H_1 H_2 T_{01}}{H_2 T_{02} + 1} = -H_1 \frac{T_{01}}{T_{02}} \tag{8.13}$$

Note: We assume $|H_2 T_{02}| \gg 1$ because, in the frequency range where the oscillation might exist, H_2 cannot be very small, and $|T_{02}|$ is large.

If the ratio of thresholds of the nonlinear links is less than 1, then H_1 is approximately equal to it; otherwise, $H_1 = 1$. Then, according to the Bode diagram for T_{01}/T_{02} in Fig. 8.22, (8.13) is satisfied if $20 \log H_1 \geq x_1$. The equation cannot be satisfied (because of the phase condition) if $20 \log H_1 < x_1$. This agrees with the experiments.

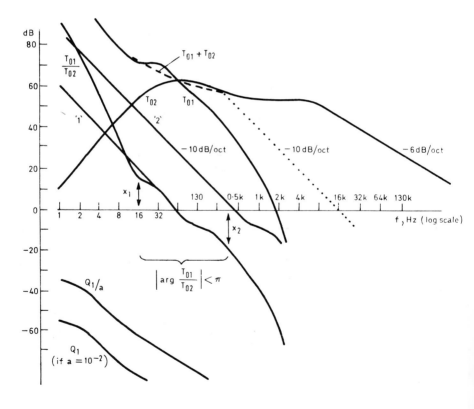

FIG. 8.22

A similar analysis can be made of the system of Fig. 8.1(a). Here, the condition of oscillation is

$$1 = -\frac{H_1 T_{01}}{H_2 T_{02} + 1} \cong -\frac{H_1}{H_2} \frac{T_{01}}{T_{02}} \qquad (8.14)$$

Approximately, H_1/H_2 equals the ratio of thresholds of the nonlinear links. When

$$20 \log |H_2/H_1| \, \epsilon \, (-x, x_1) \qquad (8.15)$$

then, the modulus of the right-hand part of (8.14) equals 1 if and only if $20 \log |T_{01}/T_{02}| \, \epsilon \, (-x, x_1)$. However, in the frequency band where this condition sustains, $|\arg T_{01}/T_{02}| < \pi$, so the condition (8.14) is not met. Therefore, condition (8.15) determines the region for H_1/H_2 for the system to be stable.

Usually, the interval $(-x, x_1)$ is much larger than the uncertainty of thresholds in nonlinear devices. Therefore, in designing the globally stable two-channel system, no special attention need be paid to the discrepancy of the nonlinear links' characteristics.

8.5.5 Experiment

In order to check the theoretical results, a physical, nonlinear, two-channel feedback system was designed and tested experimentally.

The nonlinear links are assumed to have characteristics of saturation, so they can be fully characterized by their thresholds. With these nonlinear links, the systems in Fig. 8.18(a) and in Fig. 8.18(b) are equivalent.

In the systems of Fig. 18.18, the nonlinear link of the plant represents the power output device. In order to simplify the modeling, the system of Fig. 8.18(b) can be transformed into the equivalent form of Fig. 8.19, which contains two nonlinear links that are equal to each other.

The NDC in Fig. 8.19 is divided into two parts, the preamplifier and NDC'. In NDC', the limiter and the preceding linear link comprise an active device (amplifier) with unstable parameters. The other two linear links in NC ensure the proper return ratios, T_{02} and $T_{01}/(1 + T_{02})$.

The objective of the system to be designed is to achieve $|T_{01}/T_{02}|$ more than 83 dB over the band up to 1.5 Hz. The mode of realization of the system in Fig. 8.19 is presented in Fig. 8.20, and in more explicit form in Fig. 8.21.

The lowpass Q_1 is assumed to have the frequency response shown in Fig. 8.15 and mimicked by an RC circuit (for instance, Q_1 may represent the path of the energy flow from a heater to a sensor in an industrial furnace).

The NDC' and the preamplifier are realized by active RC circuits consisting of operational amplifiers (OA). The limiter in the main channel is realized by an OA with large internal feedback. The gain and the threshold of this link can be regulated by potentiometers R_1 and R_2.

Figure 8.22 shows the Bode plots for T_{01}, T_{02}, $T_{01} + T_{02}$ (dashed curve), and T_{01}/T_{02}. The Bode plots are similar to the diagrams of Fig. 8.16.

The diagrams for T_{01}, T_{02}, and $T_{01} + T_{02}$ were calculated assuming that the low-frequency gain of the OA is large (over 90 dB) and its 0 dB gain frequency $f_T = 5$ MHz. The diagram for T_{01}/T_{02} was calculated and measured.

The average slope of the diagram for $T_{01} + T_{02}$ is less than 10 dB/octave. Then, if (8.11) holds, the system satisfies the criterion for global stability.

The diagram for T_{02} (and then for $T_{01} + T_{02}$) may have steepness up to 10 dB/octave, as shown by the dotted line in Fig. 8.22. Hence, there is some freedom in designing T_{02}, which may be used to simplify realization of the diagram for T_{01}/T_{02}.

The diagram for T_{01}/T_{02} determines the frequency response of arg T_{01}/T_{02}. For this reason, the theoretical response for $|T_{01}/T_{02}|$ must be realized with high accuracy. Since the calculated response is rather complicated, high-order approximation must be used. In the experiments, the 16-order function T_{01}/T_{02} was applied.

The variations of the gains of the OA applied in the NDC do not noticeably affect the Bode diagram for T_{01}/T_{02}; this will be shown as follows.

Because the gain and f_T of the applied OA are large, the local feedback in the preamplifier is great enough to stabilize its parameters. In NDC', the variation of the gain of the OA causes T_{01} and T_{02} to vary simultaneously. This is acceptable, since the ratio T_{01}/T_{02} will not be changed, and the Bode plot for $T_{01} + T_{02}$ will only move upward or downward as a whole, and in any such case the global stability criterion will be satisfied.

In the system under examination, the noise N was mainly produced by the OA in the preamplifier. The noise, as measured at the input of the limiter of the main channel, is shown in Fig. 8.23.

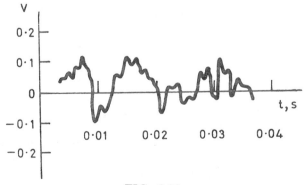

FIG. 8.23

The noise depends on the shape of the high-frequency part of the Bode diagram for T_{01}/T_{02}. The shape shown in Fig. 8.22 was selected as a good compromise between the feedback that can be attained in the band up to 1.5 Hz, on the one hand, and the values of noise and stability margin, on the other.

The frequencies of the dominant noise components belong to the interval [20, 500] Hz. According to (8.12), the amplitudes of higher-frequency components are significantly smaller.

Compare the two-channel system with the single-channel system having either Bode diagram 1 or 2 in Fig. 8.22, each with standard slope of $-10\,dB/octave$. The noise in system 1 is the same as in the two-channel system because these two diagrams coincide in the frequency band of dominant noise components. However, the feedback at 1.5 Hz in this system is only 53 dB, i.e., 30 dB smaller than the feedback in the two-channel system. The diagram 2 provides for the feedback at 1.5 Hz to be equal to that for the two-channel system (83 dB). However, in system 2 the noise components in the range above 16 Hz are about 20 dB greater than in the two-channel system. Thus, the two-channel system is appreciably better than the single-channel one.

The single-loop system with the Bode diagram for T_{01} shown in Fig. 8.22 is stable. According to the Bode criterion for the multiple-loop system, the two-channel system is stable if the Nyquist diagram for $T_{01}/(T_{02}+1) \cong T_{01}/T_{02}$ does not encircle the critical point, i.e., if $|\arg T_{01}/T_{02}| < \pi$ at a frequency where $|T_{01}/T_{02}| = 1$. To prove the stability of the system, let us consider the Bode diagram for T_{01}/T_{02}.

In the band [2, 16] Hz, the average slope of the diagram for T_{01}/T_{02} is $-20\,dB/octave$, which produces $|\arg T_{01}/T_{02}| \cong 300°$ in the middle of this frequency region. The slope of the high-frequency asymptote of this diagram is $-24\,dB/octave$, which corresponds to the phase shift $-360°$.

In the band [16, 400] Hz, the phase lag is less than 180°. The amplitude stability margins are as follows: at 16 Hz, the upper stability margin is $x_1 = 16\,dB$; at 400 Hz, the lower stability margin is 20 dB. They were measured in the following manner: by potentiometer R_1 the gain around the loop was increased until oscillation started, the increase of gain then being x; similarly, the minimum decrease of the gain that causes the system to oscillate is x_1.

Experimentally, stability of the system was examined while varying the voltage of the power supply, delivering signals of various amplitudes and shapes to the system's input, varying the values of some linear elements of the system, and combining these factors. In all cases, the system became stable when the system's parameters were returned, continuously or abruptly, to the nominal values.

In other words, the experiments have not shown that the system is *not* globally stable. Thus, the system appears to be globally stable, i.e., an oscillation cannot be excited, if and only if the ratio of thresholds of saturation in NDC to that of the main channel is less (in dB) than x_1. If this ratio exceeds x_1, the initial condition leading to oscillation can be easily found and created. The frequency of oscillation is approximately equal to the frequency at which $20 \log |T_{01}/T_{02}| = x_1$. This result agrees with the theory.

While a periodic signal (sinusoidal, square-wave, or saw-toothed) of any frequency and amplitude was applied to the input, no jumps at the output were observed. This result agrees with the criteria of Sec. 4.3.4.

The analysis and synthesis procedures were executed in the frequency domain, and the frequency optimum criterion was used: maximum feedback in the prescribed frequency band. It is of further interest to analyze the transient output responses following a step to the input.

As indicated in Sec. 2.4.2, the overshoot is predominantly determined by the phase stability margin, and the rise time is determined by the frequency at which $|T_{01}/T_{02}| = 1$. In the system under examination, these two parameters are equal to those of a conventional single-channel system whose Bode plot is shown by curve 1 in Fig. 8.22. The transient responses measured at the input and output of the lowpass Q_1/a are shown in Fig. 8.24(a).

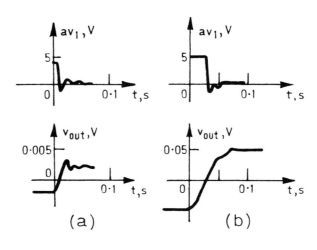

FIG. 8.24 (a, b)

The settling time is $\sim 2/1.5 \cong 1.5$ s, relative to the operational frequency band of up to 1.5 Hz. Due to increased feedback at 1.5 Hz, the accuracy achieved is considerably better than that in the single-channel system 1.

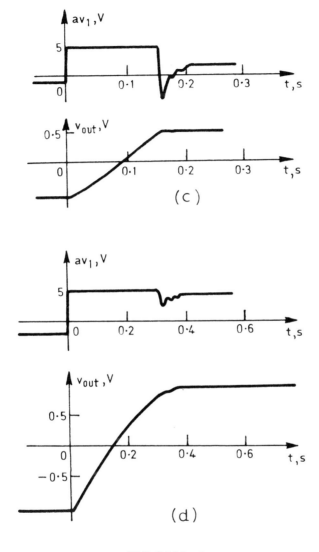

FIG. 8.24 (c, d)

Figure 8.24 shows the transient responses of the input and output of the lowpass Q_1/a, while the amplitude of the step to the input is so large that the signal is limited in the nonlinear links.

The rise time of the transient response in Fig. 8.24(b) might be reduced by decreasing the stability margins for large-amplitude signals, but this will produce large wind-up for some input signals.

In Fig. 8.19, the preamplifier was considered as a linear link; thus, precautions must be taken to preclude the saturation in the OA of the preamplifier from affecting the stability condition. The forced oscillation can be excited if the threshold of saturation in the preamplifier is lower than that in the NDC.

In the device shown in Fig. 8.21, no troubles were caused by this condition. In the slightly modified system, however, the high-level input signal whose frequency belongs to the interval [2, 10] kHz excites high-amplitude forced oscillation (subharmonics) whose fundamental is ~15 Hz. In this system, the lowpass RC filter was inserted between the output of the OA of the preamplifier and NDC′, replacing the feedback branch consisting of 0.02 μF, 2 kΩ, and 0.02 μF. The performance of the modified system was in all aspects equal to that of the system of Fig. 8.21, except for the above-mentioned subharmonics.

8.6 NDC WITH TWO NONLINEAR ELEMENTS

As previously mentioned, the application of the system in Fig. 8.2(b) is limited to only low-frequency feedback systems because of the difficulty, or even impossibility, of attaining large negative feedback in the NDC at high frequencies. At higher frequencies, the system of Fig. 8.1(a) could be of use as well as the system of Fig. 8.25, which is the combination of the two.

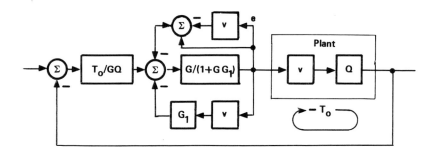

FIG. 8.25

The combined system of Fig. 8.25 may be analyzed with the scalar AS criterion, since its three nonlinear links have common input, and the system is thereby reducible to an equivalent single-channel system. However, it is easier to find the domain of the system's application with the DF approach, as will be discussed in the following.

The upper feedback path of the NDC reduces the NDC gain at lower frequencies for high-level signals. The lower feedback path, containing the highpass G_1, reduces the NDC gain at high frequencies for low-level signals, and this path becomes ineffective for high-level signals due to saturation. Both of them, together, decrease the steepness of the Bode diagram for the equivalent linear system (recall Sec. 5.2.7), thus reducing the phase lag in the critical state.

While designing such a system for wideband applications, toward the goal of implementing negative feedback in the NDC, we can utilize the excess gain, on the order of 10 dB, in the area near ω_b that normally must be leveled off while shaping the lower step on the loop-gain Bode diagram (recall Sec. 2.4.5).

The upper feedback path prevents the oscillation condition for being fulfilled at lower frequencies at a higher level of the signal; the lower path does the same at higher frequencies for low signal amplitudes.

If the negative feedback in the NDC in the neighborhood of ω_b is, for example, only 10 dB, then increasing E up to 1.5 does not make a noticeable change in $|T|$ at higher frequencies, but at lower frequencies it reduces $|T|$ by $|Q+2| \cong |Q+1|$ times. Figure 8.26 shows that $|T(1.5, j\omega)| \cong |T_0(j\omega)|$, since at higher frequencies $|T| \cong |T_0| = |T_B|$. At lower frequencies, where $|Q| \gg 1$, $|T_e| \gg 1$, $|T_0| \gg 1$, we have

$$T(1.5, j\omega) \cong \left|\frac{T_e}{1+Q}\right| = \left|\frac{1+T_0}{1+Q}\right| = |1 + T_e| \cong |T_0| = |T_B|$$

Thus, also, for $E > 1.5$ an oscillation cannot occur.

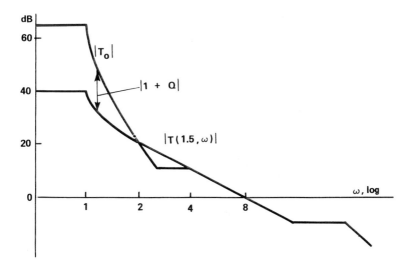

FIG. 8.26

8.7 SWITCHING REGULATION

When an inertial object is driven by a two-position relay regulator (on-off, or "bang-bang" control), there arises the problem of how to determine the exact timing of the switching in order to minimize the chosen norm of the plant's output error. If the plant dynamic is known initially, application of the Pontriagin principle solves the problem. It is noteworthy, however, that rather small deviations from this calculated optimal timing, or deviations of the plant dynamics from those used in the calculation of optimal timing, typically generate very large output errors. When the plant parameter ignorance is substantial, accurate regulation can only be achieved by using a closed-loop system with large feedback, which will "automatically" calculate and perform the switching.

The relay element of the actuator can be viewed as a saturation element preceded by an infinite gain amplifier. Therefore, even a small noise at the actuator input will cause the relay contacts to dribble. Hence, the equivalent gain of the actuator must be reduced to a finite value.

The approach commonly employed in precision system design is to modulate the relay element with a certain clock frequency, the duty cycle being proportional to the signal amplitude (pulsewidth modulation). If the clock frequency is sufficiently large, the pulsewidth modulator is equivalent to a finite gain amplifier with saturation, and the system can be designed using standard methods. Using a smaller clock frequency improves the actuator efficiency, since each switching consumes a certain amount of energy, but introduces a delay in the feedback loop, thus reducing the available feedback. Even with the lowest clock frequency ω_s permitted by the requirement of the feedback maximization ($\omega_b > 3\omega_s$), the rate of switching is much higher than that which is theoretically required for the control to be optimal [45].

In cheaper, low-accuracy regulators, the equivalent gain of the actuator is reduced by introducing hysteresis in the relay element, either through implementation of local positive feedback around the relay element, or introducing an extra element with hysteresis (Schmidt trigger) at the input of the actuator.

With small hysteresis and large gain in the linear links of the loop, comparatively small deviations of the output signal from the reference can cause the actuator to switch. In steady-state operating conditions, the actuator switches periodically, and between the switchings the error at the output of the integrating object changes exponentially between the maximal deviations on both sides of the desired value.

In other words, the accuracy depends on the equivalent low-frequency loop gain. This gain must be bounded at higher frequencies in order to reduce the noise effects of the employed sensor and amplifier. Therefore, in the switching regulator control, as well as when using a saturation-type actuator, the gain-frequency response of the linear links of the loop must be as steep as is allowed by the stability conditions and the wind-up reduction.

We can see in Fig. 3.26 that the maximum phase lag of the DF of the link with hysteresis is 90°. The larger the signal amplitude, the smaller is the phase lag and the smaller is the modulus of the described function.

The coarse DF stability analysis of the switching regulator shows that the phase lag of T_0 must be chosen as $\sim 90°$ and, correspondingly, the slope of the Bode diagram for T_0 as $\sim -6\,\mathrm{dB/octave}$. This slope is not big enough for the higher harmonics to be neglected. The total effect of higher-harmonic interaction shifts the fundamental by up to 30°, for which 30° of the stability margin in T_0 should be reserved. The rest of the total stability margin of 60° is allocated for matching the hysteresis phase lag and for reducing the forced oscillation* down to an acceptable value. Because of a shallow Bode diagram for T_0, the available feedback in this system is considerably lower than in the system using pulsewidth modulation.

Introducing a NDC chosen from among the types considered in Chapters 5, 7, and 8 at the input to the relay actuator with hysteresis allows for an improvement in the system performance, permitting the use of steeper Bode diagrams for T_0. The effect is twofold: the equivalent loop phase shift for the fundamental reduces, first, due to the action of the NDC, and, second, due to the decreased influence of higher harmonics. Such a regulator yields a simple and rather high quality implementation of closed-loop optimal control.

The nonlinear element in the NDC may be of saturation type or with hysteresis. When this element characteristic differs from the actuator nonlinearity, DF analysis may be applied with reasonable accuracy because, due to the NDC application, the slope of the Bode diagram is allowed to be steep.

---

*In relay systems, the forced oscillation produces a variety of quite different-looking wave shapes, depending on the stability margins and the type of input signal [169].

List of Symbols

$A = \operatorname{Re} \theta$

A_0 — maximum available feedback over the normalized frequency band [0,1]; value of A at $s=0$; coefficient in Laurant expansion of $\theta(\omega)$

A_∞ — value of A at $s=\infty$

A_I, A_II — constant values of A over the segments 1 and 2 of piecewise-constant frequency response $A(\omega)$

$A_c = A(\omega_c)$

A_r — reflection attenuation

A_T — asymptotic losses in feedback loop

$B = \operatorname{Im} \theta$

B_1 — coefficient in Laurant expansion of $\theta(\omega)$

$B_c = B(\omega_c)$

B_n — nonminimal phase shift

$B_{nc} = B_n(\omega_c)$

$B_u(A)$ — boundary for a diagram on the Nichols chart

C — capacitance

E — e.m.f. of the signal source; amplitude of the fundamental at the input to a loop nonlinear link

E_1 — E at the input to NDC

E_2 — E at the input of the plant

E'_a — value of E after the downward jump

E''_a — value of E after the upward jump

E'_b — value of E before the downward jump

E''_b — value of E before the upward jump

E_b — confluent E'_b and E''_b

E_c — value of E causing $|T| = 1$

$\overline{E}$ — mean square Gaussian noise e.m.f.

F — return difference

F_0 — return difference in the linear state of operation

$F(0)$ — value of F when two specified terminals are shorted

$F(\infty)$ — value of F when two specified terminals are open

$|F|$, or $20 \log |F|$ — feedback

$G(s)$, G — transmission coefficient of a linear link

H — describing function
H_1 — describing function of NDC
H_2 — describing function of actuator
K — transmission coefficient; transmission coefficient in voltage
K_E — ratio of the output voltage to the source e.m.f.
K_I — transmission coefficient in current
K_β — closed-loop transmission coefficient in voltage
$K_{E\beta}$ — ratio of the output voltage to the e.m.f. of the source in a closed-loop system
$K_{I\beta}$ — closed-loop transmission coefficient in current
L — inductance
$M = T/F$; gain coefficient of an amplifier
Q — regulation of variable equalizer; plant's linear link
R_L — load resistance
S — sensitivity
$S_m = \mathrm{Re}\, S$
$S_\phi = \mathrm{Im}\, S$
S_z — sensitivity of driving-point impedance
T — return ratio; absolute temperature
$T(0)$ — value of T when two specified nodes are connected
$T(\infty)$ — value of T when two specified nodes are disconnected
T_0 — T in linear state of operation
$T_{0\,max}$ — value of T_0 when the loop gain is maximal
$T_{0\,min}$ — value of T_0 when the loop gain is minimal
$T_B(j\omega)$ — return ratio for Bode optimal cut-off
T_c — T when $|T| = 1$
T_e — T in an equivalent system
$T_{0i} = T_0$ for the ith channel, other channels switched off
U — amplitude of a sinusoidal incident signal
U', U'' — threshold values of U causing jumps in E
V — amplitude of fundamental at the output of nonlinear link
V_{s1}, V_{s2} — V when the nonlinear link is saturated
V_{sb} — V of a subharmonic
V_k — V of kth harmonic
W — immittance, or transmission coefficient
$W(0)$, $W(\infty)$ — values of W related to zero and infinite values of regulating element w
$W = W(w_0) = \sqrt{W(0)\,W(\infty)}$ — medium value of W of symmetrical regulator
$X = \mathrm{Im}\, Z$
Y — admittance

List of Symbols

Z — impedance

Z_l, Z_n — driving point impedances with piecewise-constant frequency responses of the module and the phase

$a_c = \mathrm{Re}\, g_c$ — characteristic (image) attenuation of a two-port, i.e., attenuation when the source impedance is z_{c1} and the load impedance is z_{c2}

$b_c = \mathrm{Im}\, g_c$ — characteristic phase of a two-port

c — width in octaves of a Bode segment; index denoting the critical state of $|T| = 1$

e, $e(t)$ — signal at the input to a nonlinear link

e_d — dead zone

e_s — saturation threshold

f — frequency

f_T — frequency at which amplification coefficient is 1

g_c — characteristic (image) constant

k_1, k_2, k_1', k_2' — transmission coefficients (or immitances) in the combiner and splitter of a feedback system

k_{1n}, k_{2n} — transmission coefficients, as in above, for a noise source

k_d — coefficient of forward propagation through a feedback circuit

m — transformation coefficient

n — number of gain stages; coefficient of asymptotic slope of a Bode diagram ($-6n$ dB/octave); transformation coefficient

q — nonlinear distortion coefficient; coefficient of changes in amplitude; Popov's stability criterion constant

$s = \sigma + j\omega$ — complex operational variable

t_r — rise time

t_s — settling time

$u = \ln \omega / \omega_c$

u, $u(t)$ — signal at system's input

v, $v(t)$ — signal at the output of a nonlinear link

w — immittance, or transmission coefficient of a regulating circuit element

w_{33} — active two-pole immittance

w_{43} — amplifier transmission immittance in a feedback system

w_0 — value of w at which $W = W_0$, where $W_0 = \sqrt{W(0)\, W(\infty)}$

x — lower amplitude stability margin in dB

x_1 — upper amplitude stability margin in dB

y, or $y\pi$ — phase stability margin

y_f — value of y at the fundamental frequency of self-oscillation

y_l — y at lower frequencies

z_{c1}, z_{c2} — characteristic (image) impedances of a two-port:
  z_{c1} is the two-port input impedance when the load impedance is z_{c2};
  z_{c2} is the output impedance when the source impedance is z_{c1}

α, β, γ — coefficients of a bilinear function
β — feedback path transmission coefficient or immittance
$\beta(0)$ — β when certain specified nodes are connected together
$\beta(\infty)$ — β when certain specified nodes are disconnected
β_0 — transmission coefficient of a feedback two-port between the output splitter and the input combiner
Γ — reflection coefficient
Δ — main determinant
$\Delta°$ — value of Δ when specified circuit element equals 0
Δf — bandwidth
$\theta, \theta(s)$ — circuit function satisfying Bode's conditions
χ, χ_1, χ_2 — clipping angles
μ — transmission coefficient of an amplifier
$\phi = \arg T_0$
ϕ_n — nonminimal phase shift of T_0
$\psi = \arg H$
ψ_c — value of ψ related to the critical case of $|T| = 1$
$\omega = 2\pi f$
$\omega_a, \omega_b, \omega_c, \omega_d$ — frequencies characterizing Bode's optimal cut-off
ω_i — center frequency of a segment of constant slope of a Bode diagram
$\omega_k, \omega_l, \omega_m, \omega_n$ — frequencies at which the locus for $T_0(j\omega)$ passes the corners of the rectangle of stability margins
ω_c — current frequency at which the phase shift must be calculated; frequency of the beginning of Bode's step on the optimal Bode diagram; frequency at which $|T| = 1$
ω_s — sampling frequency
ω_{Tav} — frequency at which the gain of several cascaded amplification stages is 0 dB

Bibliography

1. J. Ackerman, "Ueber einen Zusammenhang zwischen der Beschreibungfunkzion und der Metode von V. M. Popov," *Regelungstechnick,* vol. 13, H. 11, 1965.
2. D. P. Athertone, *Nonlinear Control Engineering,* New York: Van Nostrand, 1975.
3. D. P. Athertone, *Stability of Nonlinear Systems,* New York: John Wiley and Sons, 1981.
4. M. A. Aizerman, "Plotting Resonance Curves for the System with Nonlinear Feedback," *Inzhenerny Sbornik,* vol. 13, 1952.
5. M. A. Aizerman, "On a Problem Concerning Stability in the Large of Dynamic Systems," *Usp. Mat. Nauk,* vol. 4, 1949.
6. M. A. Aizerman and F. R. Gantmacher, *Absolute Stability of Regulator Systems,* San Francisco: Holden Day, 1964.
7. B. D. O. Anderson, "Stability of Control Systems with Multiple Nonlinearities, *J. Franklin Institute,* vol. 282, 1966.
8. D. M. Auslander, Y. Takahashi, and M. Tomizuka, "Direct Digital Process Control: Practice and Algorithms for Microprocessor Applications," *Proc. IEEE,* vol. 66, no. 2, 1978, pp. 199–208.
9. K. J. Astrom (guest editor), Special Issue on Self Tuning Control, *Optimal Control: Applications and Methods,* vol. 3, no. 4, 1982.
10. V. M. Beliavtsev and B. J. Lurie, "Oscillation Modes in a Feedback Amplifier," *TUIS,* vol. 56, 1971 (in Russian).
11. V. M. Beliavtsev and I. N. Zhukov, "Feedback Maximization in a System with Parallel Channels," *Vopr. Radioelektroniki, TPS,* no. 4, 1970, pp. 23–29 (in Russian).
12. V. M. Beliavtsev and I. N. Zhukov, "Jump Resonance in a System with Parallel Channels," *Vopr. Radioelektroniki, TPS,* no. 4, 1970, pp. 118–120 (in Russian).
13. V. M. Beliavtsev and I. N. Zhukov, "Experimental Analysis of an Amplifier with Parallel Channels of Amplification," *Vopr. Radioelektroniki, TPS,* no. 3, 1971, pp. 88–91, (in Russian).
14. V. M. Beliavtsev and I. N. Zhukov, "Design of a System with Parallel Channels of Amplification," *Vopr. Radioelektroniki, TPS,* no. 3, 1971, pp. 69–75 (in Russian).

15. H. S. Black, "Stabilized Feed-Back Amplifiers," *Electrical Engineer*, vol. 3, 1934; reprinted in *Proc. IEEE*, vol. 72, no. 6, 1984.
16. J. H. Blakelock, *Automatic Control of Aircraft and Missiles*, New York: John Wiley and Sons, 1965.
17. R. B. Blackman, "Effect of Feedback on Impedance," *BSTJ*, no. 3, 1943.
18. F. H. Blecher, "Multi-Loop Transistor Feedback Amplifiers," *Proc. Nat. El. Conf.*, 1957.
19. F. H. Blecher, "Prevention of Overload Instability in Conditionally Stable Circuits," US Patent No. 2,986,707.
20. M. L. Blostein, "On the Effect of Loss in Filter Networks," *Proc. of the Third Allerton Conf. on Circuit and System Theory*, Univ. of Illinois, 1965, pp. 421–429.
21. H. W. Bode, "Variable Equalizers," *BSTJ*, April 1938.
22. H. W. Bode, "Broad Band Amplifier," US Patent No. 2.367.711.
23. H. W. Bode, *Network Analysis and Feedback Amplifier Design*, New York: Van Nostrand, 1945.
24. H. W. Bode, "Feedback — the History of an Idea" *Proc. of the Symp. on Active Network and Feedback Systems*, vol. X, 1960.
25. H. W. Bode and C. Shannon, "Simplified Derivation of Linear Least Square Smoothing and Prediction Theory," *Proc. IEEE*, vol. 38, no. 4, 1950.
26. N. N. Bogoliubov and Y. A. Mitropolsky, *Asymptotic Methods in the Theory of Nonlinear Oscillation* (English translation) New York: Gordon and Breach, 1961.
27. Z. Bonnen, "Stability of Forced Oscillation in Nonlinear Feedback System," *IEEE Trans. Automatic Control*, vol. AC-6, Dec. 1958, pp. 109–111.
28. R. W. Brockett and J. L. Willems, "Frequency Domain Stability Criteria," Parts 1 and 2, *IEEE Trans. Automatic Control*, vol. AC-10, nos. 3, 4, 1965.
29. R. W. Brockett and H. B. Lee, "Frequency Domain Instability Criteria for Time-Varying and Nonlinear Systems," *Proc. IEEE*, vol. 55, no. 5, 1967.
30. J. D. Brownlie, "Small Signal Response from dc Biased Devices" *Proc. IEE*, vol. 110, pp. 823–829, 1963.
31. J. D. Brownlie, "On the Stability Properties of Negative Impedance Converter," *IEEE Trans. Circuit Theory*, vol. CT-13, 1966, pp. 98–99.
32. J. L. Casti, *Nonlinear Systems Theory*, New York: Academic Press, 1985.

33. N. G. Chetaev, *The Stability of Motion* (English translation), New York: Pergamon Press, 1961.
34. D. W. Clarke, "Implementation of Self-Tuning Controllers," in Harris and Billings (eds.), *Self-Tuning and Adaptive Control Theory and Applications*. New York: Peter Peregrinus, 1981.
35. D. W. Clarke and P. J. Gawthorp, "Self-Tuning Controllers," *Proc. IEE*, vol. 122, no. 9, 1975.
36. D. W. Clarke and P. J. Gawthorp, "Self-Tuning Control," *Proc. IEE*, vol. 126, no. 6, 1979.
37. J. C. Clegg, "A Nonlinear Integrator for Servomechanisms," *Trans. AIEE*, Pt. II, Appl. Ind., March 1958.
38. P. A. Cook, "Circle Theorems and Functional Analysis Methods in Stability Theory," *Proc. IEE*, vol. 126, no. 8, 1979.
39. G. W. M. Coppus, S. Shah, and R. K. Wood, "Robust Multivariable Control of a Binary Distillation Column," *Proc. IEE*, vol. 130, Pt. D, no. 5, 1983.
40. E. J. Davison, "Multivariable Tuning Regulators: the Feedforward and Robust Control of a General Servomechanism Problem," *IEEE Trans. Automatic Control*, vol. AC-21, 1976.
41. P. Decauln and J. C. Gille, "Stability of Forced Oscillation in Nonlinear Systems of Automatic Control," *Proc. I Congress IFAC*, Moscow, 1960.
42. V. R. Demidovich, *Lectures on Mathematical Theory of Stability*, Moscow: Nauka Press, 1967 (in Russian).
43. C. A. Desoer and E. S. Kuh, *Basic Circuit Theory*. New York: McGraw-Hill, 1969.
44. R. Fano, "Theoretical Foundation for the Different Impedances Wide-Band Matching," *J. Franklin Inst.*, vol. 249, nos. 1, 2, 1950.
45. A. A. Feldbaum, *Optimal Control Systems* (English translation), New York: Academic Press, 1965.
46. R. Feldtkeller, "Die Berechnung der Ruckkopplungsferzerrungen bei Leitungen mit Zweidrahtverstarkern," *TFT*, no. 10, 1925.
47. P. Fleisher, "Design of Passive Adaptive Linear Feedback Systems with Varying Plants," *IRE Trans. Automatic Control*, vol. AC-7, no. 2, 1962.
48. U. Foster, D. Geeseking, and U. Weymayer, "Nonlinear Filters with Independent Transmission for the Phase and Amplitude, with Applications," *Trans. ASME, Theoretical Foundations for Engineering Calculations*, Ser. D, no. 2, 1966.
49. A. Fucuma and M. Matsubara, "Jump Resonance Criteria in Nonlinear Control Systems," *IEEE Trans. Automatic Control*, vol. AC-11, no. 4, 1966.

50. A. Fucuma and M. Matsubara, "Jump Resonance in Nonlinear Feedback Systems — Part I: Approximate Analysis by the Describing-Function Method," *IEEE Trans. Automatic Control,* vol. AC-23, no. 5, 1978.
51. E. D. Garber, "Frequency Criteria for the Absence of Periodic Responses," *Automatica Remote Control,* vol. 28, 1967.
52. P. Garnell, *Guided Weapon Control Systems,* 2nd ed. New York: Pergamon Press, 1980.
53. J. L. Garrison, L. P. Labbe, and C. C. Rock, "Basic and Regulating Repeaters," *BSTJ,* vol. 48, no. 4, 1969.
54. K. Geher, *Theory of Network Tolerances,* Budapest, Academiai Kiado', 1971.
55. A. Gelb and W. E. Vander Velde, *Multiple-Input Describing Functions and Nonlinear Systems Design,* New York: McGraw-Hill, 1968.
56. A. G. Glattfelder and W. Schaufelberger, "Stability Analysis of Single Loop Control System with Saturation and Antireset-Windup Circuits," *IEEE Trans. Automatic Control,* vol. AC-28, no. 12, 1983.
57. L. S. Goldfarb, "About Some Nonlinearities in the Regulator Systems," *Automatika i Telemechanika,* vol. 8, no. 5, 1947; translated in T. Oldenburger (ed.), *Frequency Response,* New York: Macmillan, 1956.
58. J. J. Golombesky, "A Class of Minimum Sensitivity Amplifiers," *IEEE Trans. Circuit Theory,* vol. CT-14, no. 1, 1967.
59. R. S. Graham et al., "New Group and Supergroup Terminals for L Multiplex," *BSTJ,* no. 2, 1963.
60. E. A. Guillemin, *Communication Networks,* New York: John Wiley and Sons, vol. 1, 1931; vol. 2, 1935.
61. G. E. Hanzell, *Filter Design and Calculation,* New York: Van Nostrand, 1969.
62. H. Hatanaka, "The Frequency Responses and Jump Resonance Phenomena of Nonlinear Feedback Control Systems," *Trans. ASME,* vol. D-85, June 1963, pp. 236–242.
63. Chihiro Hayashi, *Nonlinear Oscillation in Physical Systems,* New York: McGraw-Hill, 1964.
64. E. G. Holzmann, "Non-Linearity in Process Systems," *IRE Trans. Automatic Control,* vol. AC-5, 1958, pp. 3–6.
65. I. M. Horowitz, *Synthesis of Feedback Systems,* New York: Academic Press, 1963.
66. I. M. Horowitz, "Comparison of Linear Feedback Systems with Self-Oscillating Systems," *IEEE Trans. Automatic Control,* vol. AC-9, 1964, pp. 386–392.

67. I. M. Horowitz, "Optimum Loop Transfer Function in Single-Loop Minimum Phase Feedback System," *Int. J. Control*, vol. 18, 1973.
68. I. M. Horowitz, "Synthesis of Feedback Systems with Non-Linear Time Varying Uncertain Plants to Satisfy Quantative Performance Specifications," *Proc. IEEE,* vol. 64, no. 1, 1976.
69. I. M. Horowitz, "Quantative Feedback Theory," *Proc. IEE,* vol. 129, Pt. D, no. 6, 1982.
70. I. M. Horowitz, J. W. Smay, and A. Shapiro, "A Synthesis Theory for the Externally Excited Adaptive System," *Trans. IEEE Automatic Control,* vol. AC-19, no. 2, 1974, pp. 101–107.
71. I. M. Horowitz and Uri Shaked, "Superiority of Transfer Function over State-Variable Methods in Linear Time-Invariant Feedback System Design," *Trans. IEEE Automatic Control,* vol. AC-20, no. 1, 1975.
72. I. M. Horowitz and Marsel Sidi, "Synthesis of Feedback Systems with Large Plant Ignorance for Prescribed Time-Domain Tolerances," *Int. J. Control,* vol. 16, 1972.
73. S. A. Hovanessian, *Radar System Design and Analysis*, Dedham, MA: Artech House, 1984.
74. Y. S. Hung and A. G. J. McFarlane, *Multivariable Feedback: a Quasi-Classical Approach*, New York: Springer Verlag, 1982.
75. J. C. Hsu and A. U. Meyer, *Modern Control Principles and Applications,* New York: McGraw-Hill, 1970.
76. Yu. Itskis, *Control Systems of Variable Structure,* New York: John Wiley and Sons, 1976.
77. Yu. Itskis, "Dynamic Switching of Type I/Type II Structures in Tracking Servosystems," *IEEE Trans. Automatic Control*, vol. AC-28, no. 4, 1983.
78. M. Joshi, "Robustness of Velocity Feedback Controllers for Flexible Spacecraft," *IEEE Trans. Aerospace and Electronic Systems,* vol. AES-21, no. 1, 1985.
79. H. Kaizuka, "A Small Diameter Coaxial Cable System," *Japan Telecommunication Review*, vol. 4, no. 2, 1962.
80. R. E. Kalman, "Physical and Mathematical Mechanisms of Instability in Nonlinear Automatic Control Systems," *Trans. ASME,* vol. 79, no. 3, 1953.
81. R. E. Kalman, "When Is a Linear Control System Optimal?" *J. Basic Engineering,* vol. 86, March 1964; reprinted in *Telecommunication and Radio Engineering,* no. 6, 1975 [126].
82. R. E. Kalman, "Lyapunov Functions for the Problem of Lur'e in Automatic Control," *Proc. Nat. Acad. Sci.,* (US), vol. 4a, no. 2, 1963.

83. S. Karni, *Network Theory: Analysis and Synthesis.* Boston, MA: Allyn and Bacon, 1966.
84. W. D. Koenigsberg and D. C. Dunn, "Jump-Resonant Frequency Islands in Nonlinear Feedback Systems," *IEEE Trans. Automatic Control,* vol. AC-20, April 1975, pp. 208–217.
85. R. L. Kosut and B. Friedlander, "Robust Adaptive Control: Conditions for Global Stability," *IEEE Trans. Automatic Control,* vol. AC-30, no. 7, 1985.
86. M. Z. Kozlowsky, *The Nonlinear Theory of Vibration-Prevention Systems,* Moscow: Nauka Press, 1966 (in Russian).
87. H. Kwakernaak and R. Sivan, *Linear Optimal Control Systems,* New York: John Wiley and Sons, 1972.
88. L. E. Larson, R. J. Burns, M. E. Levy, and W. W. Cheng, "An Analog CMOS Autopilot," *IEEE J. of Solid-State Circuits,* vol. SC-20, no. 2, 1985.
89. C. T. Leonides (ed.), *Guidance and Control of Aerospace Vehicles,* New York: McGraw-Hill, 1966.
90. E. Levinson, "Some Saturation Phenomena in Servomechanisms with Emphasis on the Tachometer Stabilized System," *Trans. AIEE,* vol. 72, Pt. II (Applications and Industry), 1953.
91. P. S. Lindsay, "Period Doubling and Chaotic Behavior in a Driven Anharmonic Oscillator," *Phys. Rev. Lett.,* vol. 47, 1981, pp. 1349–1352.
92. F. B. Llevellyn, "Some Fundamental Properties of Transmission Systems," *Proc. IRE,* no. 3, 1952.
93. C. A. Ludeke and W. Pong. "Concerning Subharmonic Oscillation Which May Exist in Nonlinear Systems Having Odd Restoring Forces," *J. Basic Engineering, Trans. ASME,* Ser. D, vol. 81, no. 1, 1959.
94. A. I. Lurie and V. N. Postnikov, "On the Theory of Stability of Control Systems," *Prikl. Mat. i Mech.,* vol. 8, no. 3, 1944 (in Russian).
95. A. I. Lurie, *Some Nonlinear Problems in the Theory of Automatic Control,* Moscow: Gostechizdat, 1951 (in Russian).
96. B. J. Lurie, "Negative Impedances Analysis," *Proc. LEIS,* vol. 4, 1959 (in Russian).
97. B. J. Lurie, "On the Maximum Stability of Negative Impedances," *TUIS,* vol. 3, 1960 (in Russian).
98. B. J. Lurie, "A Method for Noise Reduction in Linear Amplifiers," USSR Patent No. 130068, 1960.
99. B. J. Lurie, "On the Zeros of the Sum of Two Impedances in the Right-Half Plane," *Telecommunications,* no. 9, 1962.

100. B. J. Lurie, "Amplifier of Electrical Current," USSR Patent No. 146779, 1962.
101. "Multistage Multiloop Feedback Amplifier," USSR Patent No. 158305, 1963.
102. B. J. Lurie, "Feedback in the System Incorporating Parallel Amplification Channels," *Telecommunications*, no. 12, 1964.
103. B. J. Lurie, Transistor Feedback Amplifiers Design, Moscow: Svyasizdat, 1965 (in Russian).
104. B. J. Lurie, "Maximal Feedback in a Two-Channel System with Saturation," *TUIS*, vol. 2, 1966 (in Russian).
105. B. J. Lurie, "On the Jump Resonance," *Vopr. Radioelektroniki*, vol. XI, no. 1, 1966 (in Russian).
106. B. J. Lurie, "The Second Subharmonic in a Feedback Amplifier," *Telecommunications*, no. 12, 1968.
107. B. J. Lurie, "Odd Subharmonics in a System with Saturation," *TUIS*, vol. 38, 1968 (in Russian).
108. B. J. Lurie, "Two-Channel Feedback System," *TUIS*, vol. 41, 1968 (in Russian).
109. B. J. Lurie, "Smooth Control of the Steepness of the Bode Cut-off," *TUIS*, vol. 43, 1969 (in Russian).
110. B. J. Lurie, "Instability of Forced Oscillation," *TUIS*, vol. 49, 1970 (in Russian).
111. B. J. Lurie, "On the Application of the Absolute Stability Criteria to the Analysis of Feedback Amplifiers with Maximal Feedback," *TUIS*, vol. 48, 1970 (in Russian).
112. B. J. Lurie, "Nonlinear Interstage Correction in a Feedback Amplifier," *Telecommunication and Radio Engineering*, no. 4, 1971.
113. B. J. Lurie, Feedback Maximization in Amplifiers, Moscow: Svyaz Press, 1973 (in Russian).
114. B. J. Lurie, "Non-Cascade Cofigurations of the Amplifier Elements in a Feedback Amplifier," *Telecommunications and Radio Engineering*, no. 3, 1977.
115. B. J. Lurie, "LC-Coupling in the F/O System Preamplifier," *Proc. 11th IEEE Int. Convention*, Tel-Aviv, 1979.
116. B. J. Lurie, "The Two-Channel Non-Linear Feedback System," *Int. J. Control*, vol. 31, no. 2, 1980.
117. B. J. Lurie, "The Experimental Study for Globally-Stable Nyquist-Stable Two-Channel Feedback System," *Int. J. Control*, vol. 32, no. 1, 1980.
118. B. J. Lurie, "On the Absolute Stability Criterion," *Proc. IEEE*, vol. 68, no. 10, 1980 (Lett).

119. B. J. Lurie, "Absolutely Stable Feedback System with Dynamic Nonlinear Corrector," *Proc. IEEE,* vol. 70, no. 8, 1982 (Lett).
120. B. J. Lurie, "Nonlinear Correction for Feedback Maximization, Describing Function Approach," *Proc. of the American Control Conference,* San Diego, 1984.
121. B. J. Lurie, "The Absolutely Stable Nyquist-Stable Nonlinear Feedback System Design," *Int. J. Control,* vol. 40, no. 6, 1985.
122. B. J. Lurie, "Feedback Representation of a Noisy Two-Port," *Proc. IEEE,* vol. 74, no. 1, 1986 (Lett).
123. B. J. Lurie and E. V. Zeliach, "A Method for Physical Realization of an Ideal Power Converter," *Radiotechnika i Elektronika,* vol. 5, no. 12, Acad. Sci. USSR, 1960 (in Russian).
124. B. J. Lurie and Yu. P. Osipkov, "On the Physical Realization of the Nyquist-Stable Feedback Amplifier," *Vopr. Radioelektroniki,* vol. XI, no. 9, 1970 (in Russian).
125. B. J. Lurie, V. M. Beliavtsev, and V. A. Zhuravlev, "A Method for the Bode Diagrams Measuring," *Telecommunication and Radio Engineering,* no. 6, 1975.
126. B. J. Lurie, V. M. Beliavtsev, *et al.,* "Nyquist Stable Linear Amplifier," *Telecommunication and Radio Engineering,* no. 7, 1976.
127. L. A. MacColl, *Fundamental Theory of Servomechanisms,* New York: Van Nostrand, 1945.
128. A. G. J. MacFarlane (ed.), *Frequency-Response Methods in Control Systems,* New York: IEEE Press, 1979.
129. A. G. J. MacFarlane and N. Karcanias, "Relationships Between State-Space and Frequency-Response Concepts," *Proc. Seventh IFAC World Congress,* preprints, vol. 3, 1978, pp. 1771–1779.
130. A. G. J. MacFarlane and B. Kouvartakis, "A Design Technique for Linear Multivariable Feedback Systems," *Int. J. Control,* vol. 25, no. 6, 1977.
131. T. Matsumoto, "A Chaotic Attractor for Chua's Circuit," *IEEE Trans. Circuits and Systems,* vol. CAS-31, Dec. 1984.
132. D. McDonald, "Nonlinear Technique for Improving Servo Performance," *Proc. of the Nat. El. Conf.,* Chicago, vol. 6, 1950, pp. 401–421.
133. D. J. McLean, *Broadband Feedback Amplifiers,* New York: John Wiley and Sons, 1982.
134. D. McLean and C. Eng, "Globally Stable Nonlinear Flight Control System," *Proc. IEE,* Pt.D. vol. 130, no. 3, 1983.
135. B. McMillan. "Multiple-Feedback Systems," US Patent 2.748.201.
136. D. T. McRuer, "A Feedback Theory Analysis of Airframe Cross-Coupling Dynamics," *J. Aeronaut. Sci,* vol. 29, 1962, pp. 525–533.

137. A. I. Mees, *Dynamics of Feedback Systems,* New York: John Wiley and Sons, 1981.
138. A. I. Mees and D. P. Athertone,"The Popov Criterion for Multiple Loop Feedback Systems," *IEEE Trans., Automatic Control,* vol. AC-25, no. 5, 1980.
139. R. G. Meyer, R. Eshenbach, and W. M. Edgerly, Jr., "A Wide-Band Feedforward Amplifier," *IEEE Trans. Solid-State Circuits,* vol. SC-9, no. 6, 1974.
140. E. F. Mishchenko and N. H. Rosov, *Differential Equations with Small Parameter and Relaxational Oscillation,* Moscow: Nauka Press, 1975 (in Russian).
141. G. H. Mulligan, Jr., "Signal Transmission in Nonreciprocal Systems," *Proc. of the Symp. on Active Networks and Feedback Systems,* vol. X, 1960.
142. G. J. Murphy, "A Frequency Domain Stability Chart for Nonlinear Feedback Systems," *IEEE Trans. Automatic Control,* vol. AC-2, no. 6, 1967.
143. K. S. Narendra and J. H. Taylor, *Frequency Domain Criteria for Absolute Stability,* New York: Academic Press, 1973.
144. P. Naslin, *The Dynamics of Linear and Non-Linear Systems,* New York: Gordon and Breach, 1965.
145. Z. Nehari, *Introduction to Complex Analysis*, rev. ed., Boston: Allyn and Bacon, 1968.
146. R. A. Nelepin (ed), *Methods of Nonlinear Automatic Control Systems Analysis,* Moscow: Nauka Press, 1975 (in Russian).
147. G. C. Newton, Jr., L. A. Gould, and J. F. Kaiser, *Analytical Design of Linear Feedback Controls,* New York: John Wiley and Sons, 1957.
148. J. Nielsen, *Missile Aerodynamics,* New York: McGraw-Hill, 1960.
149. D. E. Norton, "High Dynamic Range Transistor Amplifiers Using Lossless Feedback," *Microwave Journal,* no. 5, 1976.
150. K. Ogata, "An Analytical Method for Finding the Closed-Loop Frequency Response of Nonlinear Feedback-Control Systems," *Trans. AIEE,* vol. 76, 1957.
151. J. Oizumi and M. Kimura, "Design of Conditionally Stable Feedback Systems," *IRE Trans. Circuit Theory,* vol. CT-4, no. 3, 1957.
152. E. Peterson, J. P. Kreer, and L. A. Ware, "Regeneration Theory and Experiment," *Proc. IRE,* Oct. 1934.
153. C. L. Phillips, H.T. Nagle, Jr., *Digital Control System Analysis and Design,* Englewood Cliffs, NJ: Prentice Hall, 1984.
154. V. A. Pliss, "On the Aizerman Problem for a System of Three Differential Equations," *Dokl. Akad. Nauk USSR,* vol. 121, 1958 (in Russian), p.3.

155. L. S. Pontriagin, *Ordinary Differential Equations* (English translation), Reading MA: Addison-Wesley, 1962.
156. H. Rothe and W. Dahlke, "Theory of Noisy Four Poles," *Proc. IRE,* vol. 44, June 1956.
157. H. H. Rosenbrock and P. Cook, "Stability and the Eigenvalues of $G(s)$," *Int. J. Control,* vol. 21, no. 1, 1975.
158. E. N. Rosenwasser, *Oscillation of Nonlinear Systems, Integral Equations Method,* Moscow: Nauka Press, 1969 (in Russian).
159. I. W. Sandberg, "A Perspective on System Theory," *IEEE Trans. Circuits and Systems,* vol. CAS-31, no. 1, 1984.
160. S. S. Sastry and C. A. Desoer, "Jump Behavior of Circuits and Systems," *IEEE Trans. Circuits and Systems,* vol. CAS-28, no. 12, 1983.
161. P. K. Sinha, *Multivariable Control: An Introduction,* New York: Marcel Dekker, 1984.
162. O. J. M. Smith, *Feedback Control Systems,* New York: McGraw-Hill, 1958.
163. A. P. Stern, "Stability and Power Gain of Tuned Transistors," *Proc. IRE,* vol. 45, no. 3, 1957.
164. V. A. Taran, "Survey: Application of Nonlinear Correction and Variable Structure for Improving Dynamics of the Automatica Control Systems," *Automatica Remote Control,* no. 1, 1964.
165. G. J. Thaler and M. P. Pastel, *Analysis and Design of Nonlinear Feedback Control Systems.* New York: McGraw-Hill, 1962.
166. H. W. Thomas, D. J. Sandoz, and M. Thompson, "New Desaturation Strategy for Digital PID Controllers," *Proc. IEE,* Pt.D, vol. 130, no. 4, 1983.
167. Yu. I. Topcheev (ed.), *Nonlinear Compensators in Automatic Control Systems,* Moscow: Mashinostroenie Press, 1971.
168. J. G. Truxal, *Automatic Control System Synthesis,* New York: McGraw-Hill, 1955.
169. J. Z. Tsypkin, *Theory of Relay Systems of Automatic Control,* Moscow: Gostechizdat, 1955. English Translation: Ya. Z. Tsypkin, *Relay Control Systems,* New York: Cambridge University Press, 1984.
170. J. Z. Tsypkin, "Absolute Stability and the Process Absolute Stability in Nonlinear Sample-Data Automatic Control Systems," *Automatica Remote Control,* vol. 24, no. 12, 1963, pp. 4–12.
171. A. M. Uttley and P. H. Hammond, "Stabilization of On-Off Controlled Servomechanisms," in *Automatic and Manual Control,* papers contributed to the conference in Cranfield, London, 1952, pp. 449–456.

172. M. Vidyasagar, *Nonlinear Stability Analysis,* Englewood Cliffs, NJ: Prentice-Hall, 1979.
173. J. C. West, *Analytical Technique for Non-Linear Control Systems,* New York: Van Nostrand, 1961.
174. J. C. West and J. L. Dauce, "The Mechanism of Sub-Harmonic Generation in a Feedback System," *Proc. IEE,* vol. 102, Pt. B, no. 5, 1955.
175. J. L. Wyatt, L. O. Chua, J. W. Gannett, I. C. Goknar, and D. N. Green, "Energy Concepts in the State-Space Theory of Nonlinear n-Ports: Part I — Passivity," *IEEE Trans. Circuits and Systems,* vol. CAS-28, no. 1, 1981.
176. J. J. Zaalberg van Zelst, "Stabilized Amplifiers," *Phillips Tech. Review,* vol. 9, no. 1, 1947.
177. M. Zak, "Two Types of Chaos in Non-Linear Mechanics," *Int. J. of Non-Linear Mechanics,* vol. 20, no. 4, 1985.
178. G. Zames, "On the Input-Output Stability of Time-Varying Nonlinear Feedback Systems," Pt. I and II, *IEEE Trans. Automatic Control,* vol. AC-11, April 1966; reprinted in *Frequency-Response Methods in Control Systems*, New York: IEEE Press, 1979 [128].
179. Guo-Qum Zhong and F. Ayrom, "Periodics and Chaos in Chua's Circuit," *IEEE Trans. Circuits and Systems,* vol. SAC-32, May 1985.

Index

Amplitude-amplitude-phase characteristic (AAPC), 206
Amplitude-phase characteristic (APC), 206
Asymptotic losses, 102
Aizerman, M.A., 130, 177

Balanced bridge, 14
Black, H.S., 237
Blackman's formula, 7
Bode, H.W., x, xi, *passim*
Bode's optimal cut-off, 88
Bode variable equalizers, 32

Characteristic exponent (index), 163
Charactreictic (image) parameters, 33, 311
Chebyshev filters, 66
Clegg integrator, 158
Closed loop transfer function, 2
Conjecture of a filter, 153
Conjecture of resonance, 153
Constraints for crossover frequency, 91
Crossover frequency, xi
Cross-sectioned feedback path, 6

Describing function (DF), 152
Dropping section, 161
Duffing's equation, 168

Feedback
  compound, parallel, series, 8
  positive, negative, large, 2
Feed-forward, 237

Horowitz, I.M., 55, 94, 99
Hysteresis, 154, 158

Ideal power converter (IPC), 22–23, 40
Integral of imaginary part, 68
Integral of real part, 61

Jump-resonance, 166, 176

Kalman, R., 66, 130
Kimura, M., 199
Kozlovsky, M.Z., 177

Levinson, E., 177
Liapunov, A.M., xii, 130, 159
Lurie, A.I., 129–130

Minimum phase function (MP), 51

Nonlinear dynamic compensator (NDC), 203
Negative impedance converter (NIC), 22, 56
Nyquist stable system, 199

Oizumi, J., 199
Oscillation
  multifrequency, 143
  nonperiodic, 142
  periodic, 142

Parallel feedback channels, 117, 242, 253
Phase-gain relationship, 74
Phase locked loop (PLL), 198
Piecewise-linear approximation, 75, 77
Plant, 3
Popov criterion, 130

Reflection coefficients, 27, 52
Return difference, 1
Return ratio, 1

Saturation
  combined with a dead-zone, 157
  dynamic, 157
  nondynamic, expressions for the DF, 16
  with frequency-dependent threshold, 157
Sensitivity
  of module, phase, 25
  of driving-point impedance, 20
System
  multiloop, 233
  multi-input/multi-output (MIMO), 233
  single-input/single-output (SISO), 233
  single-loop, 233

Stability
  absolute (AS), 129
  absolute of processes (APS), 170
  asymptotic global (AGS), xii
  Bode-Nyquist criterion, 234
  conditional, xii
  margins, 50
  Nyquist type, xii, 99
Subharmonics
  odd, 191
  second-order, 191
Substantial jump-resonance, 180

Wind-up, 175